CALCAREOUS ALGAE

Further Titles in this Series:

1. A.J. BOUCOT

   EVOLUTION AND EXTINCTION RATE CONTROLS

2. W.A. BERGGREN AND J.A. VAN COUVERING

   THE LATE NEOGENE

   Biostratigraphy, geochronology and paleoclimatology of the last 15 million years in marine and continental sequences

3. L.J. SALOP

PRECAMBRIAN OF THE NORTHERN HEMISPHERE

Developments in Palaeontology and Stratigraphy, 4

CALCAREOUS ALGAE

John L. Wray

ELSEVIER SCIENTIFIC PUBLISHING COMPANY

Amsterdam, Oxford, New York, 1977

ELSEVIER SCIENTIFIC PUBLISHING COMPANY
335 Jan van Galenstraat
P.O. Box 211, Amsterdam, The Netherlands

Distributors for the United States and Canada:

ELSEVIER/NORTH-HOLLAND INC.
52, Vanderbilt Avenue
New York, N.Y. 10017

Library of Congress Cataloging in Publication Data

Wray, John Lee, 1925-
Calcareous algae.

(Developments in paleontology and stratigraphy v. 4)
Bibliography
Includes index.
1. Algae, Fossil. 2. Rocks, Carbonate.
I. Title. II. Series.
QE955.W69 561'.93 76-48656
ISBN 0-444-41536-X

with 170 illustrations and 8 tables

Printed in The Netherlands

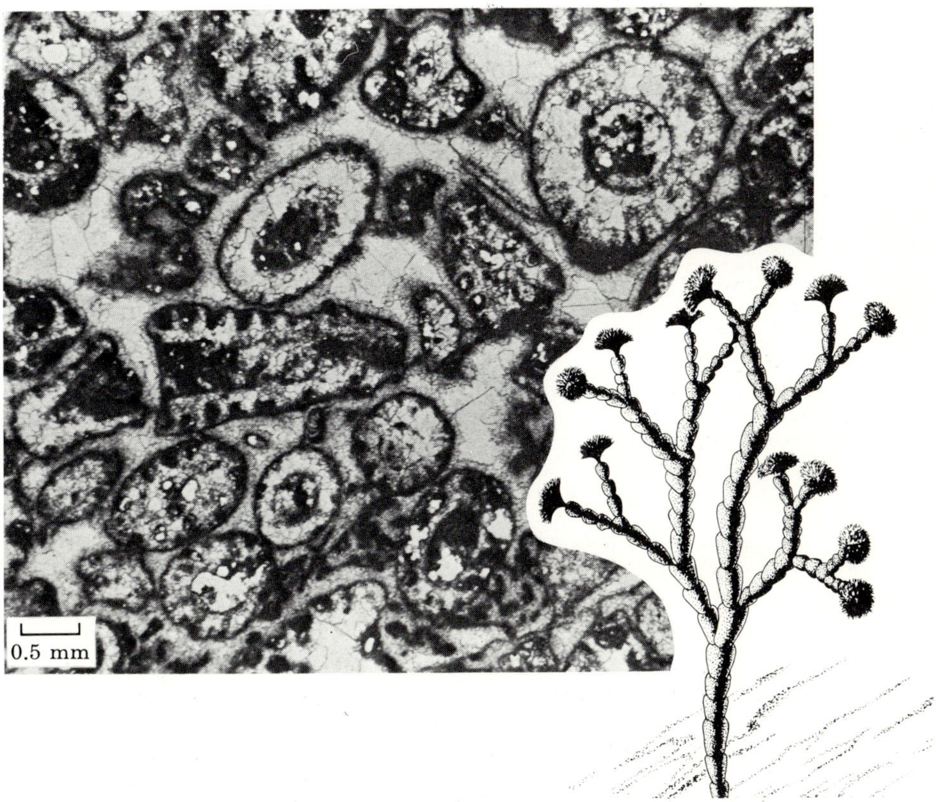

Thin section of Paleogene marine limestone from Guatemala composed of skeletal remains of calcareous green algae (Dasycladaceae) and sketch of analogous living sediment-producing plant.

In the sea and in fresh water, as well as in damp places and even in situations subject to drought, algae occur in abundance in all parts of the world.

A. C. Seward, 1898

PREFACE

Calcareous algae, by any measure, comprise a specialized subject among the many divisions of the earth sciences. Yet as a group, the algae have been ubiquitous throughout much of the earth's recorded history, some have exerted an important influence in sedimentary processes, and the skeletal remains of calcareous forms are abundant and widespread in sediments of many types and ages. Calcareous algae are significant in paleontology as records of ancient life, and they are useful in biostratigraphy and paleoenvironmental reconstructions. Furthermore, an understanding of calcareous algae is essential to comprehending modern approaches in carbonate sedimentology.

Today calcareous algae have a curious position in geology. They are readily acknowledged as common kinds of fossils, yet hardly ever considered in any detail in paleontology courses, and otherwise are generally ignored in earth science curricula, except in some sedimentology courses. Their treatment has not been commensurate with the state of knowledge, and as a result, fossil calcareous algae are poorly understood by students and professional geologists alike.

This book is written for geologists -- paleontologists, petrologists, sedimentologists, and stratigraphers -- and deals mainly with the geology of calcareous algae, including living forms and the Recent record. There is now a large body of information on calcareous algae, some of it the result of significant advances in the field of carbonate sedimentology during the last two decades, but nowhere does this appear in a single volume. My purpose here is to provide a synthesis of our knowledge of calcareous algae in the form of a comprehensive introduction; the book is not an inventory of all known calcareous algae and algal carbonates. The contents divide themselves rather naturally into two parts. The first is a systematic discussion of representative skeletal calcareous algae and algal-laminated sedimentary structures; the second considers the role of algae in the broader context of applied geology -- sedimentology, paleoecology, and biostratigraphy -- including economic applications in petroleum and mineral exploration.

The paleontological usage has been kept at a simplified level. My intent is not to discourage a rigorous taxonomy of fossil algae, but rather to make the information more useful to a larger audience. Similarly, I have passed over details of the biology of calcareous algae that are not pertinent to the identification of fossil taxa and the interpretation of paleoenvironments. More applicable biological data would be a welcomed addition to understanding the geology of calcareous algae but, unfortunately, the study of these calcareous plants has been neglected by botanists for various reasons and quantitative ecological studies are scarce.

There are many problems and unanswered questions in the field of calcareous algae, and in a synthesis of this kind areas that need further investigation become readily apparent. Fossil calcareous algae frequently have been misinterpreted throughout the history of their study, and all too often the group has served as a "wastebasket" for various unidentifiable biotic and nonbiotic constituents presumed to be algae. In addition, we still have an incomplete knowledge of the age and spatial distribution of some common forms, and the biological affinities of many are seriously questioned.

Despite these problems and the general lack of dissemination of known information, many geologists are enthusiastic about the opportunities calcareous algae seem to offer for better understanding the rocks in which they occur. I hope this book can supply some of the information and also the proper perspective for appreciating the importance, as well as the limitations, of utilizing calcareous algae in deciphering earth history.

Professor J. Harlan Johnson provided the foundation for this book. As a result of nearly 40 years of effort, he contributed a tremendous amount to our understanding of fossil calcareous algae and algal limestones. His published works provide a wealth of data and illustrations which will remain fundamental references for generations. Harlan Johnson introduced me to this subject in 1956 and for many years aided me in my studies of these fossils.

This book is the result of assistance and encouragement from many people. I am especially indebted to all of those who contributed information and suggestions to its contents. Among them are Walter H. Adey, Graham R. Davies, Robert N. Ginsburg, Alan S. Horowitz, Noel P. James, Kenji Konishi, Claude L. V. Monty, Matthew H. Nitecki, Phillip E. Playford, James F. Read, Richard Rezak, Robert Riding, Johannes H. Schroeder, and Donald F. Toomey. The Colorado School of Mines permitted me to reproduce illustrations from its publications.

Thomas R. Fisher prepared the majority of the original line drawings and reconstructions of calcareous algae, and I am indebted to him for his artistic talents and scientific understanding. Most of the graphic illustrations were done by Barbara A. Steele; others were completed by Sally M. Andrews and Richard R. Nervig. M. P. Moore did much of the photography. William H. Lohman took the scanning electron micrographs.

Bettye C. Hart typed the manuscript drafts and final camera-ready copy of the entire book. Her conscientious efforts are gratefully acknowledged. The author thanks the management of the Denver Research Center, Marathon Oil Company for providing the creative atmosphere necessary to accomplish this task and for permission to publish the results.

Finally, because this book is produced by a photographic offset process, the author alone is responsible for any errors.

July 1976 John L. Wray
Littleton, Colorado

CONTENTS

Chapter 1

INTRODUCTION

Algae are a large and diversified assemblage of aquatic, nonvascular, chlorophyll-bearing plants, ranging from microscopic cells a few microns in size to enormous benthonic seaweeds tens of meters in length. This simple definition overlooks the biological problems of adequately delimiting the algae as a group, but nevertheless characterizes most forms. Within this huge complex of plants the *calcareous algae* constitute a highly artificial group that includes:

- Various kinds of benthonic and planktonic algae whose thalli contain biochemically precipitated calcareous skeletal material.
- Mechanically accumulated deposits of calcium carbonate caused by algae -- usually an interaction of biological and physical processes -- called stromatolites and algal-laminated sediments.

Algae with calcareous skeletal matter are scattered fortuitously among several major taxa, along with noncalcareous forms, and show considerable variation in morphology and in the kind and degree of calcification. The mineralogy and extent of calcification are major factors in fossilization. Calcareous algae have had a long and continuous chronologic record and include some of the oldest fossils known. They have been remarkably successful in adapting to many different kinds of marine and nonmarine environments, but their remains are preserved principally in carbonate sediments. Studies of fossil calcareous algae require interdisciplinary approaches, dealing with biology and paleontology on the one hand and sedimentology on the other. Thus, to begin with, we need to appreciate calcareous algae as a heterogeneous complex of skeletal materials and biosedimentary structures broadly distributed in time and space.

HISTORICAL REVIEW

Living marine calcareous algae attracted the attention of curious observers hundred of years ago, although the earliest published references may be lost in antiquity. Several of the more common forms were the objects of European scientific descriptions by the early 1600s. For example, the first record of the calcareous green alga *Halimeda* (under the name of *Sertolara*) appeared in *Dell' Historia Naturale* by Imperato published in Naples in 1599 along with a short description and an excellent figure of the plant (Barton, 1901). Sloane (1707) included descriptions of various coralline red algae and *Halimeda* in his discussion of the natural history of Jamaica, and Ellis (1755) described a group of coralline algae and other forms from the British Isles. According to Migula (1897), a figure of the freshwater calcareous alga *Chara* was first published in 1623, although it was considered to be a form of *Equisetum*.

Most 18th century workers, including Linnaeus who prepared encyclopedia descriptions of the known world fauna and flora in the mid-1700s, concluded that marine skeletal calcareous algae belonged to the animal kingdom. They assumed that the presence of calcareous substances precluded the possibility of classifying them as plants.

The French naturalist Lamouroux's classical works of 1812 and 1816 established many of the common generic names recognized today, including *Amphiroa*, *Corallina*, *Jania*, *Halimeda*, *Udotea*, *Acetabularia*, *Cymopolia*, and *Neomeris*. Lamouroux and others still insisted that these calcified organisms were not plants but belonged to the corals or coral-like animal groups. An English translation of selections of Lamouroux's work published in 1824 provides an indication of the popular interest in the subject at that time. In the translator's preface we read that the purpose of the volume is to provide a handbook for those "interested in Nature's history, but unskilled in learned language". This book presumably contains the descriptions of all of Lamouroux's species (624 species assigned to 57 genera; many of them illustrated on 19 plates of figures) of various calcareous red and green algae, in addition to coelenterates and sponges.

The botanical affinity of the crustose coralline red algae was established by Philippi in 1837 whose generic names *Lithothamnium* and *Lithophyllum* are still used. The first definite record of the inclusion of the calcareous green alga *Halimeda* in the plant kingdom is obscure, but Barton (1901) believes it may have been in an unpublished manuscript of Targioni-Tozzetti in 1819. The vegetable nature of this genus, along with the calcareous green algae *Udotea* and *Acetabularia* and the articulated coralline genus *Corallina*, was confirmed later by Link (1834). Thus by the middle of the last century European naturalists had developed a rather good appreciation of the general nature of living skeletal calcareous algae and their place in the scheme of life. Man in less sophisticated societies had drawn similar conclusions. The early Hawaiians, for example, recognized *Halimeda* and the articulated corallines as "limu" or seaweeds, although the crustose coralline algae lack Hawaiian names -- presumably because they were not considered plants or animals (Soegiarto, 1973).

Some of the earliest descriptions of fossil calcareous algae seem to be those of Eocene calcareous green algae, *Ovulites* (Lamarck, 1816) and *Acicularia* (d'Archiac, 1843), from the Paris basin. These fossils were placed originally in various divisions of the animal kingdom, but later Munier-Chalmas (1879) proved them to be algae similar to living genera. The importance of crustose coralline red algae as skeletal constituents in the Miocene Leithakalk of the Vienna basin was recognized by Unger in 1858. In 1878 Nicolson and Etheridge described some minute tubular structures in Ordovician nodules from Scotland which they named *Girvanella*. The biological affinity of this rather common fossil has always been uncertain, but it generally has been classified as blue-green algae.

In the 1890s and early 1900s Rothpletz made substantial contributions to the knowledge of fossil calcareous algae by describing and interpreting diverse forms of various ages from European localities. Noteworthy is his work on crustose coralline algae. Rothpletz (1891) used reproductive organs for distinguishing fossil coralline taxa, criteria that subsequently were applied to living forms and are now a fundamental part of classification schemes. Foslie (works published between 1881-1909), who was a botanist contemporary of Rothpletz, produced numerous taxonomic studies of living crustose coralline red algae from Norway and other regions of the world.

Seward's four-volume classic *Fossil Plants*, vol. 1 (1898), provides an overview of the state-of-the-art in the field of fossil algae at the end of the 19th century. This work is remarkable because it contains much more information on fossil calcareous algae than most current paleobotany books. Of the approximately 80 pages Seward devoted to fossil algae, about one-half deals with calcareous forms. His appreciation of the problems of identifying fossil algae and the role of calcareous algae in sedimentology is illustrated by the following:

> It has been justly said that palaeontologists have been in the habit of referring to algae such impressions or markings on rocks as cannot well be included in any other group. "A fossil alga", has often been the *dernier ressort* of the doubtful student. (Seward, 1898, p. 139)

> There are ... limestones which wholly or in part owe their formation to masses of calcareous algae, which grew in the form of submarine banks or on coral reefs. Occasionally the remains of these algae are clearly preserved, but frequently all signs of plant structure have been obliterated. (Seward, 1898, p. 26)

In 1913 Garwood discussed the importance of fossil calcareous algae, especially the part they played as "rock-builders". Although he was not involved directly in very many studies of calcareous algae, his influence was instrumental in furthering an interest in the subject among geologists and paleontologists. Eighteen years later, Garwood (1931), the president of the Geological Society of London, presented an address to the Society summarizing important additions to the knowledge of fossil calcareous algae during this period.

Mme. Paul Lemoine, whose works span half a century (1909-1958), added much to our understanding of the structure, classification, and the geologic and geographic distribution of crustose coralline algae. Her studies in Paris dealt with both fossil and living forms, and she emphasized their role in sedimentology. Adey and Macintyre (1973) estimate that Lemoine produced 75 percent of the papers on crustose corallines between 1910-1940.

During about the same interval of time, Pia in Austria and Lucien and Jean Morellet in France carried out extensive studies of various fossil calcareous algae, but especially dasycladacean green algae. Pia's treatise on fossil dasycladaceans published in 1920 and a later paleobotany text (Pia, 1926) contain excellent reconstructions of dasycladacean plants which have been widely reproduced.

Beginning in the 1930s, major studies of fossil calcareous algae expanded from Western Europe to the U.S.A. and the U.S.S.R. Two individuals in particular, Johnson in the United States and Maslov in the Soviet Union, left a very productive record extending into the late 1960s. Johnson described many fossil taxa representing nearly all major groups. He also prepared numerous catalogs and review articles on the subject and three annotated bibliographies (Johnson, 1943, 1957, 1967). Johnson's *Limestone-building algae and algal limestones* (1961) represents the most comprehensive English-language treatment of the taxonomy of fossil calcareous algae. Like Johnson, Maslov's studies dealt with diverse groups of calcareous algae and include two modern comprehensive works (Maslov, 1956; Maslov et al., 1963).

The decades of the fifties and sixties witnessed a tremendous increase in studies of ancient and recent carbonate sediments which continue at a high level today in research and applied fields. This in turn generated a heightened concern for a better understanding of skeletal and nonskeletal algae in carbonate sedimentology. Much of the overall effort in carbonate rocks during this period was stimulated directly or indirectly by the petroleum industry because of the interest in limestones and dolomites as hydrocarbon reservoirs.

The evolution of thought on biosedimentary structures caused by algae -- stromatolites and algal-laminated sediments -- began much later and has proceeded along different lines. Various laminated calcareous structures in rocks, presumed to be the result of biogenic growth, were first described by Hall (1883) from the Cambrian of New York and considered to be of animal origin. This kind of structure was given the name *stromatolith* by Kalkowski (1908). Walcott (1914) concluded that Proterozoic stromatolites in Montana were comparable to freshwater calcareous tufa formed by blue-green algae. Working on Recent sediments in the Bahamas, Black (1933) was the first to recognize stromatolites as products of the interaction of mechanical sedimentation and mats of blue-green algae. Ginsburg (1955) expanded Black's studies to south Florida and directed attention to Recent sediment-binding algae as analogues of fossil stromatolites. Logan (1961) described the unique columnar stromatolites in Shark Bay, Western Australia which provided an additional contribution to an understanding of the various kinds of Recent forms. Studies of ancient and modern stromatolites and other kinds of algal-laminated sediments mushroomed in the 1960s and early seventies and numerous publications have appeared on aspects of their biology, environmental significance, and biostratigraphy.

Today work on skeletal calcareous algae and stromatolites is being pursued by paleontologists, sedimentologists, and biologists in many parts of the world. Our knowledge of living skeletal calcareous algae still suffers from inadequate treatment by botanists. An exception however, is the contemporary work of Adey who since 1964 has produced important additions to the taxonomy and distribution of Recent crustose coralline red algae. His studies on the ecology of North Atlantic crustose corallines are unsurpassed in scope and detail and they provide some fundamental parameters with which to interpret the environmental distribution of fossil coralline algae.

Taxonomic studies of fossil calcareous algae continue to be fragmentary, carried on largely by isolated specialists, and often lack unification with broader aspects of paleontology and sedimentology. Notwithstanding these difficulties, it is possible to sort through the accumulated volume of knowledge from different sources and to assemble an intelligible picture of calcareous algae and their use in deciphering geologic processes and earth history.

CONCEPTS AND DEFINITIONS

At the outset it is imperative to distinguish between three genetic categories of calcareous algae:

- Skeletal calcareous algae.
- Laminated calcareous structures (stromatolites) caused mainly by nonskeletal algae.
- Freshwater calcareous tufa.

Through usage over many years the term "calcareous algae" has been applied to representatives of all three groups, yet the distinctions, and the significance of the differences, have not always been appreciated.

Skeletal calcareous algae comprise a taxonomically diverse assemblage of algae that deposit calcium carbonate as a result of metabolic and biochemical processes. Skeletal calcium carbonate may occur in various habits and concentrations within and upon the entire plant body or may be localized in only a portion of the plant. Skeletal materials alone, that is, separated from vegetative structures of the whole plant, may represent direct or indirect evidence of the thallus, internal tissues, or various organs. Many fossil remains of skeletal calcareous algae are identifiable biological entities related to living taxa; consequently, most fossil forms are subject to the rules of botanical nomenclature and can be handled within the frameworks of existing classification schemes. On the other hand, the biological affinity of some fossil skeletal calcareous algae is uncertain because they lack preserved diagnostic features and cannot be related to modern descendants or morphological counterparts.

Laminated calcareous structures attributed to algal activity are caused primarily by the mechanical accumulation of fine-grained carbonate sediment on organic films or mats generated by colonies of nonskeletal filamentous and coccoid algae. Periodic interaction of nonskeletal algal mats and physical sedimentation produces distinctively laminated biosedimentary structures. The term *stromatolite* is applied to forms with pronounced vertical relief, whereas forms with flat-lying laminae are generally called *algal-laminated sediments*. Stromatolites have been classified on both biological and nonbiological bases. Most workers now agree that stromatolites cannot be classified biologically because they are principally sedimentary structures, but that they are amenable to empirical groupings based on geometric parameters.

Freshwater calcareous tufa is formed by nonskeletal blue-green algae. Tufa nodules and encrusting growths in lakes and streams are the result of in situ precipitation of calcium carbonate, presumably as a product of algal photosynthesis. The concentric laminations of algal tufa resemble internal features of algal stromatolites, but tufa material is generally more dense because of its biochemical origin.

These three categories of calcareous algae are sufficiently distinct morphologically and genetically to separate them in most instances. Skeletal algae and algal-laminated structures are quantitatively significant in ancient sediments, whereas tufa deposits are rare and environmentally restricted.

Skeletal and nonskeletal algae may occur in intimate association. In these cases it may be difficult to determine the relative importance of biogenic growth of skeletal algae and sediment-binding effects of nonskeletal algal mats in the development of the overall structure. Also, a few fossil skeletal algae (e.g., *Sphaerocodium*) constructed colonial growth forms that closely resemble columnar stromatolites formed by nonskeletal algae. Finally, tufa, stromatolites, and other algal-laminated sediments may look like inorganically formed structures, such as caliche (a calcareous soil deposit formed in the vadose zone), travertine, beach rock, and other inorganic chemical deposits of calcium carbonate due mainly to evaporation. Consequently, it may be virtually impossible to distinguish these kinds of structures from algal structures without a good understanding of the geology of the deposit.

The paleontological approach to skeletal calcareous algae has often been less than "systematic". By reason of their botanical affinity, it would seem appropriate to consider them in the broader context of paleobotany, yet few traditional paleobotanists -- those who concentrate on vascular plants -- concern themselves with these fossils. Therefore the taxonomy of fossil calcareous algae has been left largely to the invertebrate paleontologist, who quite often is a specialist of benthonic foraminifers. There is some logic here, because most preserved skeletal calcareous algae occur in marine limestones associated with invertebrate faunas and are microscopic constituents studied mainly in thin sections. The serious student of fossil calcareous algae is generally self-taught, combining a knowledge of the petrography of various kinds of biotic constituents and the morphology of living algae. Unfortunately, there are many invalid fossil taxa in the literature -- including both nonalgal and nonbiogenic constituents -- and this clouded situation tends to dilute the usefulness of real genera and species for geologists unfamiliar with these problems. Lists of described fossil skeletal calcareous algae may be divided into "knowns", those identifiable taxa based on analogous living forms, and "problematical algae", objects presumed to be algae but which lack diagnostic morphological criteria of modern taxa. Although problematical algae cannot be classified biologically, some can be characterized and identified readily on the basis of unique features, therefore they are useful in biostratigraphy and in biofacies analysis.

The vast majority of skeletal calcareous algae are marine plants, and so their fossil remains are preserved in marine sediments. The principal exception is the class Charophyceae, which today lives in fresh and brackish water habitats and presumably occupied similar nonmarine environments in the past. In contrast to skeletal calcareous algae, stromatolites and algal-laminated sediments occur in both marine and nonmarine settings, although they tend to be concentrated in shallow marginal waters of marine basins and saline lakes.

METHODS

Fossil calcareous algae are generally studied in petrographic thin sections. Examination in transmitted light at magnifications of about 10 to 150 times reveals gross shape and microscopic skeletal details. Most benthonic calcareous algae are macroscopic plants ranging from several millimeters to tens of centimeters in overall size, yet they are usually treated as microfossils, because they disaggregate into small segments or break up into fine particles in the depositional environment. The scanning electron microscope (SEM) is useful for examining surface features and internal structures of living skeletal calcareous algae. This instrument provides new insight into the minute structural elements of some skeletal algae, but has not yet contributed significantly to a better understanding of their taxonomy. The SEM is of limited value in studying fossil calcareous algae embedded in indurated rocks and specimens subjected to diagenetic alteration.

Traditionally, studies of skeletal calcareous algae have included only the benthonic forms, thus excluding marine planktonic calcareous algae, the minute coccolithophorids. This division continues because of radically different analytical procedures and interpretative ends. The purpose of most investigations of benthonic calcareous algae is in connection with petrographic and paleoenvironmental interpretations of carbonate rock facies, whereas coccolithophorids are used principally for biostratigraphic zonation, notwithstanding the volumetrically significant "rock-building" contribution of coccoliths to Cretaceous and Cenozoic chalks.

NONCALCAREOUS FOSSIL ALGAE

There are many kinds of *noncalcareous* fossil algae not considered here. These include boring algae, several major phytoplankton groups, such as siliceous diatoms and the dinoflagellates which are composed of resistant organic compounds, in addition to occasional records of fleshy megascopic forms belonging to the brown, red, and green algae which occur as imprints and carbonaceous films. Noncalcareous filamentous and coccoid forms preserved in cherts or diagenetically replaced have been reported from Precambrian and Phanerozoic sediments. Noteworthy occurrences are in middle and late Precambrian stromatolitic cherts (see for example Barghoorn and Tyler, 1965; Schopf, 1968) and in Devonian sediments (Fairchild et al., 1973; Wicander and Schopf, 1974).

Boring and endolithic microorganisms belong to several major groups of algae, as well as fungi. These inhabit a variety of carbonate substrates in Recent environments, such as shells and other skeletal grains, and have been recognized in fossil materials. Boring algae may be useful as paleoecological indicators (Golubic et al., 1975), and it is important to appreciate their destructive effect on carbonate grains and influence on diagenesis.

Noncalcareous algae presumed to be fossils have been recovered as organic filaments from insoluble residues of ancient limestones (de Meijer, 1969; Coron and Textoris, 1974). Whether these remains belong to sediment-building or boring algae was not established. Also, there may be some question about the antiquity of these kinds of plant remains because they could be Recent endolithic algae which have invaded the outcrop.

GEOLOGY OF CALCAREOUS ALGAE

The geologic roles of calcareous algae are many and diverse, representing a broad spectrum of direct and indirect considerations ranging from systematic paleontology to applied sedimentology. Within the realm of paleontology, there are three main areas of interest and application:

- Systematic paleontology: inquiries into the taxonomy, classification, evolution, and functional morphology of skeletal calcareous algae.
- Biostratigraphy: the use of distinctive taxa, with restricted chronologic ranges, to date and correlate rock strata.
- Paleoecology: the use of skeletal algae and algal-laminated structures, with limited environmental tolerances and spatial distributions, to interpret sedimentary environments and to characterize biofacies.

Calcareous algae are of greatest value in the reconstruction of ancient carbonate environments. They provide an important complement to other benthonic organisms in delineating marine biofacies because the group is widespread and abundant in shelf environments, yet many individual forms are restricted to narrow ecologic niches. Most benthonic skeletal algae have long geologic time ranges; consequently, they have less value for age determination and stratigraphic zonation, although a few define marker horizons in parts of the sedimentary column. Then too, for the very reasons benthonic algae are good environmental indices, they are not useful for regional stratigraphic correlations.

A knowledge of calcareous algae may be applied to carbonate sedimentology in many ways including the following:

- Carbonate sediment production: skeletal calcareous algae have been quantitatively important producers of coarse-grained sediments and sand-, silt-, and clay-sized carbonate particles in ancient and modern times, accounting for the bulk of fine-grained sediments in some environments.
- Algae-sediment interactions: these include several different sorts of relationships, such as baffling, trapping, and binding of particulate sediment by various kinds of algal mats and upright-growing forms of algae. The most significant process has been the periodic interaction of mucilaginous algal mats and physical sedimentation, resulting in the development of stromatolites and other algal-laminated sedimentary structures.
- Skeletal frameworks: some skeletal calcareous algae, for example Cenozoic crustose coralline red algae, have erected rigid, self-supporting frameworks and effectively cemented together other biotic constituents, notably corals, to construct organic buildups,

such as reefs and banks. A variation of this process is the development of hard bottoms or pavements by encrusting corallines.

- Petroleum reservoirs and occurrence of metallic minerals: algal-rich facies have provided porous and permeable reservoirs for hydrocarbon accumulations of various ages and, in a few instances, have served as preferential host rocks for ore mineralization.

Chapter 2

SKELETAL CALCAREOUS ALGAE

Carbonate skeletons in many algae serve as supportive, protective, or framework functions, but in others the material appears to be extraneous to the organism's vital activities or highly localized in the plant. Nonetheless, the term "skeletal" is a valid and useful way of distinguishing algae with calcareous "hardparts" -- species that produce sedimentary constituents and often are preserved as fossils -- from ones that do not secrete calcium carbonate under normal conditions. Algae with biochemically precipitated deposits of calcium carbonate comprise an artificial biological assemblage that cuts across taxonomic boundaries, yet they make up a thoroughly practical association in geology.

In this chapter we shall consider general aspects of skeletal calcareous algae -- morphology and growth, nature and causes of calcification, and classification. The purpose here is to introduce characteristics common to most skeletal algae before describing in detail the various systematic groups.

MORPHOLOGY AND GROWTH

Skeletal calcareous algae display a tremendous variation in morphology, ranging from very simple shapes to exceedingly complex arrangements of external and internal skeletal elements. In fact, these kinds of algae exhibit a far greater diversity of form than exists among the higher plants and within most invertebrate groups.

There are few biological treatises that discuss calcareous algae in any detail. The basic work in the field of algae by Fritch (1935, 1945) is a massive two-volume compilation of the world's literature on the morphology and reproduction of this plant group. Although this work is obviously deficient in modern advances in the physiology and classification of algae, it remains a practical and inclusive reference. Dawson (1966) prepared an up-to-date textbook overview of the physiology, ecology, and geographic distribution of marine algae. This book furnishes a useful biological perspective for the geologist because it covers aspects of benthonic calcareous algae in the context of the total flora.

Biologically, the algae have these characteristics. All algae live in water or moist environments. They carry on photosynthesis (thus require sunlight) and therefore are able to manufacture their own food [=autotrophic]. All algae contain chlorophyll *a*, but they are so distinctly colored by other pigments that color is an important basis for their classification. They evolve oxygen during photosynthesis and require oxygen for respiration, consequently differing from anaerobic photosynthetic bacteria and resembling the higher plants. The plant body of an alga -- called a thallus -- lacks conducting tissues [=nonvascular] and is not differentiated into true roots, stems, and leaves. With a few exceptions, nearly all cells within a plant can carry on photosynthesis. Beyond these common attributes the algae make up a heterogeneous group with extreme variations in size, morphology, cellular organization, reproduction, and biochemistry. The algae include both of the fundamentally different patterns of cellular organization -- the primitive prokaryotes and the specialized eukaryotes. Reproduction in the prokaryotic blue-green algae (Cyanophyta) is vegetative. In all other algae -- which are eukaryotic -- reproductive processes are vegetative, asexual, and sexual.

The thalli of living algae may consist of various internal organizations, including single cells, colonies of single cells, unbranched filaments, branched filaments, more than one kind of filament, and cellular tissue. Most phyla or divisions of algae exhibit a wide range of growth habits and internal structure, but lesser taxa of skeletal algae show a more limited range of diversity.

For purposes of discussion the morphology of skeletal calcareous algae is treated empirically in terms of:

- Growth form and external geometry.
- Internal structure.
- Skeletal microstructure.

Both external form and preserved internal tissues or other structural elements serve to characterize individual taxa. Skeletal microstructure (sometimes called ultrastructure) refers to the texture or fabric of the minerals that form the skeletal wall (Horowitz and Potter, 1971) and is less important in systematic identifications.

Growth form and external geometry

The variation in growth habits of skeletal algae is represented by these examples -- upright-growing, branched forms which look like higher plants; plain crusts; and simple microscopic filaments. The shape of some skeletal remains in the sedimentary record may reflect very little of the overall growth habit of the parent plant, because of the disarticulation or disintegration of the original skeleton. Then too, there is the inherent problem in the petrography of all skeletal material, namely the visualization of three-dimensional objects from two-dimensional views seen in thin sections. Some fossil algae are known only from skeletal fragments or disarticulated segments, thus making it difficult to reconstruct their growth habits, particularly forms lacking living analogues.

Most of the variation in external form of skeletal calcareous algae can be included in three main categories:

- Encrusting habits.
- Erect plants.
- Filamentous habits.

Encrusting skeletal algae include smooth and irregular crusts, nodules, and rigid branched forms (Fig. 1). They range in size from a few millimeters to several centimeters. Representatives of this shape category are found in various families of the red algae.

Fig. 1. Encrusting growth habits. Left to right: irregular crust, nodule, and rigid branched form.

Skeletal algae with an erect growth habit are commonly segmented, articulated, and branched plants several centimeters in height (Fig. 2). These kinds of algae almost always disarticulate into individual segments whose external geometry may be in the form of plates, cylinders, and other shapes less than a millimeter to a few millimeters in size. Erect growth habits occur principally among the articulated coralline red algae, the codiacean green algae -- *Halimeda* is a well known example -- and the dasycladacean green algae.

Skeletal calcareous algae with filamentous habits are found mainly in the blue-green algae (Fig. 3). Filament diameters usually measure a few tens of microns. It is important to note, however, that the internal structure of many skeletal algae consists of various kinds of filaments of approximately the same size, thus in poorly preserved material they could be confused with filamentous growth habits.

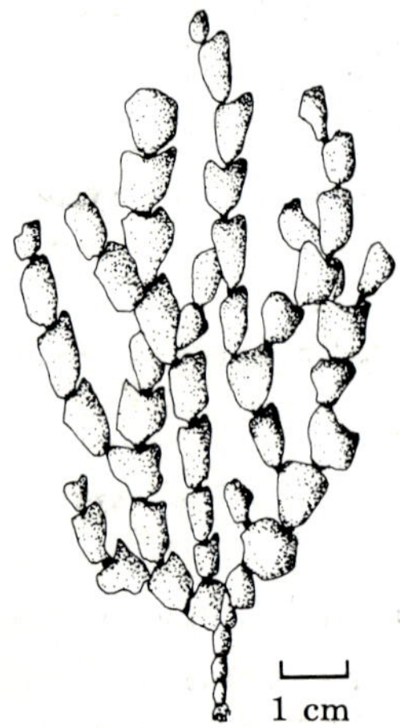

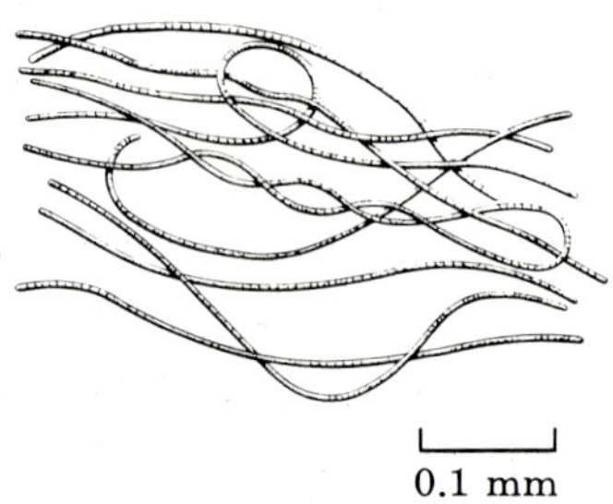

Fig. 2. Erect growth habit of codiacean green alga *Halimeda*.

Fig. 3. Filamentous growth habit of blue-green algae.

Internal structure

The geometry (shape, size, and arrangement) of various internal skeletal elements plays a major part in the identification of fossil skeletal algae, and the reason thin sections are required for detailed studies. Obviously, any alteration or obliteration of internal structure by diagenetic processes complicates identification.

Filaments are basic anatomical structures of many skeletal calcareous algae. They occur in a wide variety of forms and arrangements within the thallus, thereby producing very distinctive kinds of internal structures.

A principal kind of internal morphology is illustrated by closely packed, heavily calcified filaments of cells that form regular patterns of cellular tissue in various families of the red algae (Fig. 4). The tissue usually is differentiated into distinct types of cellular arrangements characteristic

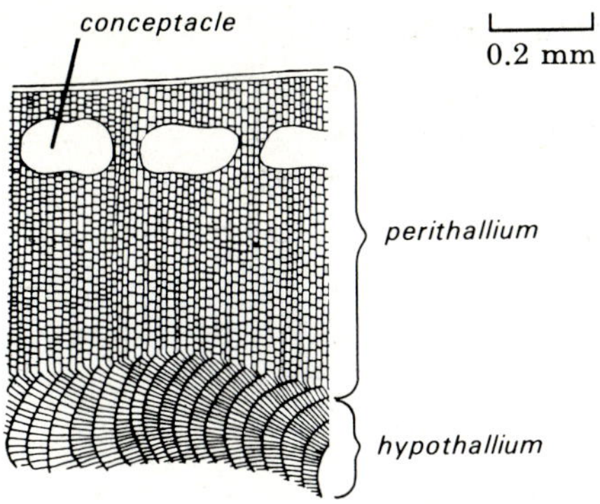

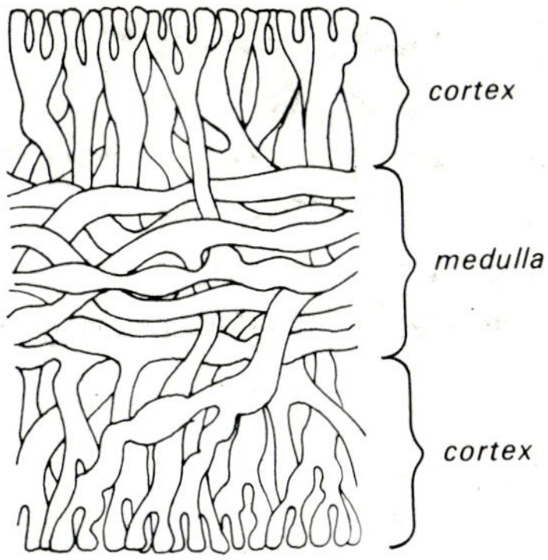

Fig. 4. Section of differentiated cellular tissue and reproductory organs in coralline algae (After Pia, 1926).

Fig. 5. Section of medullary filaments and cortex in codiacean green alga *Halimeda*.

of certain positions within the thallus. Individual cells are commonly rectangular and measure a few tens of microns.

Another kind of internal organization of filaments is typified by the green alga *Halimeda*. Here the internal structure is characterized by loosely packed, unsegmented filaments forming an outer or cortical zone of lateral branches that develop from a multiaxial core of central (medullary) filaments (Fig. 5). Filament diameters range from several tens of microns to about 100 microns. Tubular filaments of similar scale occur within the calcified reproductive organs (oogonia or gyrogonites) of the nonmarine Charophyceae.

The skeletal internal structure found in segmented erect plants belonging to the dasycladacean green algae represent a separate category not comparable genetically to the two previous types, because calcification occurs only on external surfaces of stems and branches rather than on filaments within the thallus. These skeletal algae appear typically as perforated cylinders and spheres whose internal organization consists of calcareous molds of an axial stem and several generations of radiating branches (Fig. 6). The diameter of individual branches varies from less than a tenth of a millimeter to several millimeters.

Reproductive organs of various shapes, sizes, and arrangements constitute a distinctive kind of internal structure in some skeletal calcareous algae. These may be embedded in skeletal tissue as in the coralline red algae (Fig. 4) or preserved in calcified molds as for example, the dasycladacean green algae (Fig. 6).

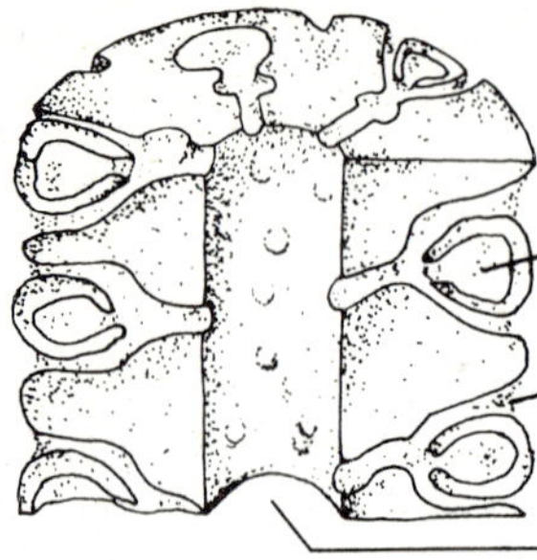

Fig. 6. Calcareous mold of dasycladacean green alga *Cymopolia*.

Skeletal microstructure

Skeletal elements in calcareous algae are composed of crystals of calcium carbonate. The morphology of individual crystals and crystal aggregates comprising the skeletal microstructure is not usually a factor in defining minor taxa, but it may be important in characterizing major groups of skeletal algae. The mineralogies of many skeletal algae are unstable and subject to diagenetic alteration, and so in these cases little or none of the original skeletal microstructure is preserved in the ancient record.

Petrographically, the skeletal microstructure of fossil calcareous algae generally appears either dark-colored or clear. Skeletons composed of very fine-grained crystals show up dark in thin section due to the repeated interference of light transmission at multiple crystal boundaries. On the other hand, larger crystals of sparry calcite are clear due to little or no interference. Fossil coralline red algae are composed of fine-grained calcite, whereas fossil calcareous codiacean and dasycladacean green algae typically have a sparry appearance in thin section as a result of the alteration of original aragonitic skeletons.

Transmission and scanning electron microscopy are useful techniques for discerning the skeletal microstructure of living benthonic calcareous algae. Studies to date have been limited to a relatively few cursory, rather than systematic, investigations of codiacean green algae and coralline red algae. Eventually, scanning electron microscopy may reveal taxonomically significant details of wall microstructure.

Individual skeletal elements of *Halimeda* and related codiacean green algae are composed of single crystals and crystal aggregates of aragonite that develop mainly in spaces between filaments (Lowenstam, 1955; Wilbur et al., 1969; Marszalek, 1971). Acicular aragonite crystals in *Halimeda* measure about 0.1-0.5 microns in width and up to a few microns in length (Fig. 7).

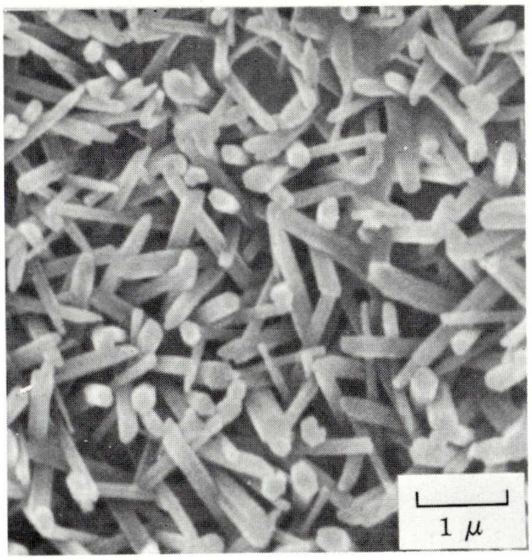

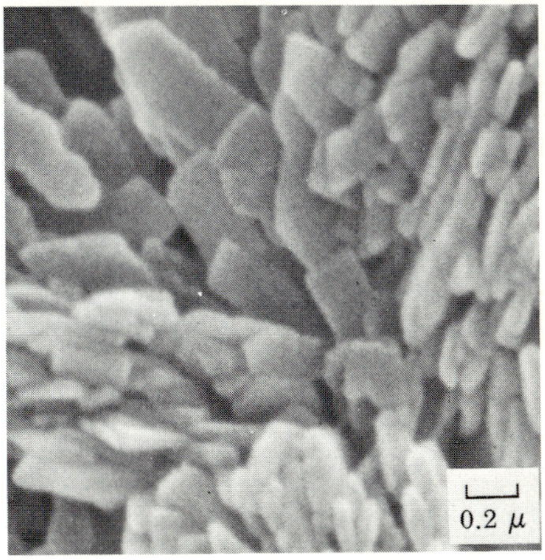

Fig. 7. Scanning electron micrograph of skeletal microstructure in codiacean green algae. Elongate crystals of aragonite in *Halimeda*.

Fig. 8. Calcitic skeletal microstructure of articulated coralline red algae. Scanning electron micrograph.

The wall microstructure of crustose coralline algae appears to be composed of blocky crystals of calcite whose long dimension is normal to the cell wall. The articulated coralline genus illustrated in Fig. 8 exhibits distinct platy or lath-like crystals approximately 0.1 micron thick and up to 0.4 micron long. Over 40 years ago Baas-Becking and Galliher (1931) investigated the wall structure of the articulated coralline alga *Corallina* and considered the causes of calcification. Using X-ray methods and the petrographic microscope, they concluded that the wall is composed of individual calcite crystals "a few tenths of a micron long ... arranged with the c-axis perpendicular to the longitudinal axis of the fiber (= filament)" -- findings confirmed with modern electron microscopes.

Growth rates

Very little direct and quantitative data exist with regard to the growth rates of skeletal calcareous algae, although some generalizations and orders of magnitude can be established. Techniques for measuring rates of growth have differed substantially and have not always been very precise. Biologists have stressed the importance of the standing crop and organic productivity in relation to growth, while sedimentologists have emphasized sediment production.

Inherent physiological differences between species accounts for the wide range of growth rates observed among the skeletal calcareous algae -- some plants simply grow faster than others in a given environment. Goreau (1963) noted that calcification rates vary according to the stage of growth and development of a species or population. In general, skeletal codiacean green algae have some of the highest growth rates, whereas crustose coralline red algae have distinctly slower rates. Numerous physical environmental factors -- temperature, light intensity and quality, circulation, salinity, nutrients -- in addition to various outside chemical and biological controls, influence rates of growth. However, the purpose here is to describe the range and variation of growth rates; the effects of environmental factors on the growth and distribution of algae are dealt with in Chapter 7 (Calcareous algae and their environments).

The growth rates of the crustose coralline red algae have been reported from various field and laboratory observations. Lemoine (1940) found that shallow water crustose species of *Lithophyllum* and *Phymatolithon* along the French coast added material at rates of 2-7 mm per year. Huvé (1954) reported new plants of *Lithophyllum tortuosum* in the western Mediterranean attained diameters of 1-2 cm in about eight months. Observations of crustose corallines in the Gulf of Maine by Adey (1965) show growth rate variations ranging from 3 mm per year for *Clathromorphum circumscriptum* at shallow depths to 0.2-0.3 mm per year for *Clathromorphum compactum* in deeper waters. Adey and McKibbon (1970) observed tagged in situ specimens of *Lithothamnium* species off the west coast of Spain over a 10-month period and noted highly seasonal growth. Most of the growth took place during June and July -- up to 5 microns per day -- with little or no growth during the winter months.

Adey (1970) carried out laboratory experiments on six species of Norwegian crustose corallines in natural sea water tanks under controlled temperature and light conditions to determine growth rates and the effects of environmental factors. Utilizing "summer-winter" light values and temperature ranges comparable to the natural environment, a mean yearly growth of about 2.5 mm per year was obtained for *Clathromorphum circumscriptum*. These laboratory rates correspond with results from field studies in the Gulf of Maine on the same species. The highest growth rates -- over 14 microns per day -- were recorded for *Phymatolithon polymorphum* (Fig. 9). Adey noted that because rapid growth provides a competitive advantage for available substrate space, the high growth rates of *Phymatolithon polymorphum* and *Lithothamnium glaciale* observed in the laboratory may explain why these species are the most abundant crustose corallines in the North Atlantic.

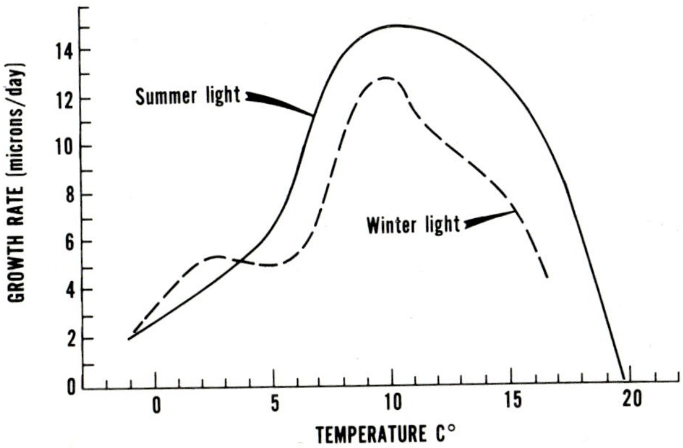

Fig. 9. Laboratory measurements of marginal growth rate of crustose coralline alga *Phymatolithon polymorphum* as a function of temperature and light. (After Adey, 1970).

Later studies (Adey and Vassar, 1975) of tropical crustose corallines (*Neogoniolithon*) recorded marginal growth rates of 0.9-2.3 mm per month, which are about an order of magnitude greater than those previously measured in subarctic waters.

Crustose corallines in deep water may have exceedingly slow rates of growth. Adey and Macintyre (1973) estimated that a continuously growing rhodolith (a laminated algal nodule of crustose corallines) of 20-30 cm diameter would be expected to be at least 500-800 years old. Considering occasional burial, exhumation, and periodic damage by browsing animals, actual ages could be expected to be much higher.

Articulated coralline red algae grow more rapidly than crustose forms according to available data. Johnson (1961) cites examples of unnamed genera of articulated corallines from the west coast of France whose growth rates averaged 3-4 cm per year when observed over about a four-year period. Johansen (1969) studied stages in the growth and development of the articulated coralline genus *Calliarthron* on anchored surfaces off the Monterey Peninsula in California and found that plants approximately one-year old had attained a height of 2-3 cm.

There is little quantitative data on the growth rates of the common tropical calcareous green alga *Halimeda*. Barton (1901) notes simply that a plant grown under observation at Funafuti Atoll in the Pacific increased in length several inches in six weeks. More recent data obtained in laboratory cultures by Culinvaux et al. (1965) provide a better understanding of

Halimeda growth rates, but their observations did not cover the entire life span of a plant. The initial growth of young sprouts amounted to approximately 4 cm in 36 days. In more mature plants an interesting pattern was observed in which new segments developed daily or every other day for one or two weeks, then no segments formed for about a month. It generally has been observed that in their natural environment *Halimeda* plants several tens of centimeters high mature in a single growing season, although significant growth-rate differences among species might be expected because the genus inhabits a wide range of environments.

The codiacean green alga *Penicillus* has a rapid growth rate and a short life span. In laboratory cultures, Culinvaux et al. (1965) observed that plants attained a height of several centimeters at maturity in 10-14 days after the tip of the stalk first appeared above the substrate. Stockman et al. (1967) documented the rates of growth of *Penicillus* in south Florida in their sedimentological investigation of lime mud production by algae. Based on the surveillance of several stations in different settings over a one-year period, the life span of *Penicillus* was observed to vary from 30-63 days. The plant produces from six to nine crops per year in the areas investigated and grows in large numbers, thus it is an important source of fine-grained carbonate sediment in these environments.

CALCIFICATION

The relative abundance of skeletal calcareous algae in the total world-wide flora is only a few percent; consequently, we conclude that calcification in the algae is a rare phenonenon. Calcified species almost always make up a minor component of any given population of living benthonic marine and nonmarine algae, although the actual proportion varies with the environment.

Utilizing data from Dawson (1966), it is estimated that about 8 percent of the approximately 8,000 living species of marine benthonic algae are calcified to any extent, and these belong mainly to a few families of the red and green algae (Table I). The proportion of calcified species among nonmarine algae is much less; they are found only in the Charophyceae and in certain groups of blue-green algae in specialized environments, such as thermal springs.

The relative diversity of marine calcareous algae compared to noncalcareous species may be appraised more realistically in terms of carbonate sedimentology by examining floristic data obtained by Croley and Dawes

TABLE I

DISTRIBUTION AND ABUNDANCE OF CALCAREOUS SPECIES AMONG MAJOR PHYLA OF MARINE BENTHONIC ALGAE.

	Approx. No. Total Species	*Approx. No. & Percent Calcareous Species*	
GREEN ALGAE	900	90	10%
BROWN ALGAE	1,500	2	0.1%
RED ALGAE	4,000	590	15%
BLUE-GREEN ALGAE	1,500	-	-
Total:	7,900	682	8%

(1970) in a shallow-water carbonate environment in south Florida. They assembled a species census (exclusive of blue-green algae) over an 18-month period along five 300-meter transects extending from the littoral zone to depths of about 10 meters. Table II lists the relative abundance of calcified species in the total flora based on their checklist of 259 benthonic species. Approximately 11 percent of the flora consists of heavily calcified species -- forms with identifiable skeletal components that likely would be preserved in the sedimentary record. An additional 9 percent are lightly calcified species whose sedimentary contributions are unidentifiable, very fine-grained, calcium carbonate particles. Thus, in this Recent carbonate environment about 20 percent of the algal flora -- just the benthonic, macroscopic forms -- are calcareous species analogous to the kinds of algae that have produced skeletal constituents in the geologic past.

TABLE II

RELATIVE ABUNDANCE OF BENTHONIC CALCAREOUS ALGAE, CONTENT KEYS, FLORIDA (COMPILED FROM CROLEY AND DAWES, 1970).

		Calcified Species			
		Heavy		*Light*	
	*Total Species**	*n*	*(%)*	*n*	*(%)*
GREEN ALGAE	78	11	(14)	8	(10)
BROWN ALGAE	29	--	--	2	(7)
RED ALGAE	152	17	(11)	13	(9)
Total:	259	28	(11)	23	(9)

*Exclusive of blue-green algae.

Nature and origin of skeletal carbonates

More is known about the nature and location of skeletal carbonates in algae than is understood about their function and precipitation mechanisms. Calcium carbonate is found within the cell, associated with cell walls, outside the cell as extracellular material, and as surficial deposits on the thallus (Fig. 10). The extent of calcification ranges from slight and loosely bound extraneous material to extensive deposition that is an integral part of the thallus.

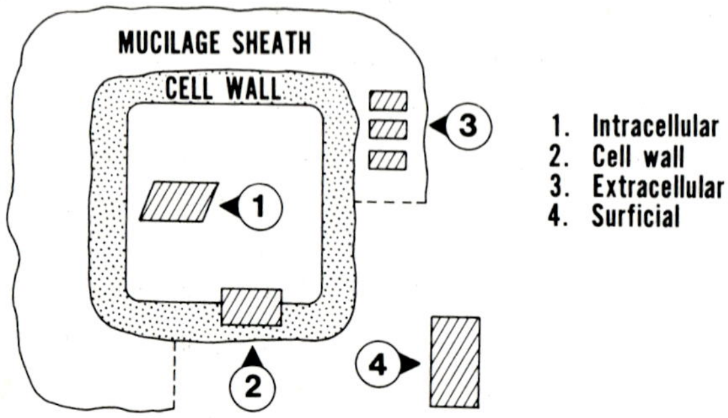

Fig. 10. Diagrammatic sketch of calcium carbonate deposition sites in algae. (After Arnott and Pautard, 1970).

In general, internal calcification is related to cellular morphology, metabolism, and the selective absorption or assimilation of specific carbonate salts. Little-understood, complex reactions of these factors result in carbonate precipitation and distribution within the thallus. The calcium carbonate is a consequence of the interaction of carbonate, or biocarbonate, with calcium under alkaline conditions, but the process by which carbon dioxide is released or absorbed to achieve this result is not clear. Pobequin (1954) suggested that the cell provides only the chemical environment in which the crystalline phase develops, and that the physical shape of the mineral product may be controlled by other factors.

Intracellular calcium carbonate is found in single-celled algae. Examples include calcite coccoliths formed by the planktonic Coccolithophoraceae and aragonite skeletal elements in the dasycladacean alga *Acetabularia*, a biologically unique plant with only a single large nucleus for most of its life. From the standpoint of morphological variation and amount of skeletal material produced, coccolithophorids are the principal kinds of algae that manufacture skeletal carbonates within the cell. The description of calcification in *Coccolithus huxleyi* by Wilbur and Watabe (1963), and a later summary of calcification in unicellular organisms by Pautard (1970), provide some understanding of the nature and origin of intracellular skeletal carbonate. Coccoliths are believed to form as a result of a special cellular system specifically for that purpose and are not the by-products of other cell processes. A coccolith originates within a membrane-bound vacuole in which calcification begins at the margins and spreads throughout the connecting membrane to form the final fully developed coccolith.

Precipitation of calcium carbonate in association with the cell wall is best exhibited in the red algae. Bailey and Bisalputra (1970) suggested that Golgi-derived vesicles in the cell wall of articulated coralline red algae are involved in the calcification process. Calcification is initiated in the outer part of the wall and subsequently extends inwards, leaving an uncalcified lamella surrounding the protoplast.

Extracellular skeletal material, including surficial deposits, are the most common types of calcium carbonate precipitation in marine algae. They occur in various red algae, in codiacean and dasycladacean green algae, and in some blue-green algae in specialized environments.

Much of the calcification in the codiacean green algae occurs outside the cell wall but within the thallus. Electron microscope studies of *Halimeda* and related codiaceans (Wilbur et al., 1969; Marszalek, 1971) show that calcification begins in the intracellular spaces beneath the cortical cells. With continued calcification the intracellular areas throughout the thallus may become filled with acicular crystals and crystal aggregates. The regular size of the crystals and their orientation suggest some internal ordering mechanisms, but it is unknown to what extent the cell controls crystallization. More recently, a second mode of precipitation, intracellular calcification, has been discovered in the codiacean genus *Penicillus* where individual aragonite crystals are contained entirely within the cytoplasm (Perkins et al., 1972).

In contrast to the seemingly complex and little understood mechanisms of internal carbonate precipitation in algae, external or surficial precipitation of calcium carbonate is believed to be primarily a product of the extraction of carbon dioxide from the water by the plant during photosynthesis and related to the chemistry of the surrounding waters.

The solubility of calcium carbonate increases with a rise in the partial pressure of carbon dioxide, decreasing temperature, and increasing amounts of sodium chloride in solution. Consequently, in a marine environment of normal salinity, maximum precipitation of calcium carbonate can be expected in the tropical, lower latitudes where warmer water temperatures and a lower carbon dioxide partial pressure are found. Generally, the external precipitation of carbonate by algae, especially codiaceans and dasycladaceans, is favored in these more advantageous environments. Thus it has been widely assumed that carbonate deposition is a direct result of photosynthesis, namely, shifting the equilibrium of the following equation towards the right:

$$Ca^{++} + 2HCO_3^- \leftrightarrow CaCO_3 + H_2O + CO_2$$

Lewin (1967) pointed out that one might therefore expect all photosynthetic marine algae to be coated with calcium carbonate, and consequently, "the problem is not so much why only certain algae deposit calcium carbonate but why most seaweeds *do not* do so."

This problem has not yet been solved. No doubt photosynthesis is an integral part of the process of extracellular precipitation, but it is not necessarily the initiating or controlling factor in calcification.

Mineralogy

Skeletal calcium carbonate in living algae is either calcite or aragonite, never mixtures of the two minerals in the same alga. This is unlike molluscan skeletons and some other invertebrates in which more than one mineral species occur together.

Red algae deposit both aragonite and magnesium calcites, the latter with varying amounts of magnesium ranging up to 30 mole percent, calculated as magnesium carbonate in solid solution in the calcite (Chave, 1952). The skeletal elements in marine green algae are aragonite exclusively. Calcified reproductive organs of the nonmarine Charophyceae are calcite. The coccolithophores deposit a low-magnesium calcite. Calcification in blue-green algae is limited to nonmarine forms and is mostly calcite, although the chemistry of the local environment is a major factor in determining the mineralogy. A single genus of marine brown algae, *Padina*, forms surficial deposits of aragonite, but these are not known from the fossil record. Table III summarizes the groups of algae that deposit skeletal calcium carbonate.

The amount of magnesium in the skeletal calcite of some crustose coralline red algae varied seasonally in response to water temperature fluctuations. Studies of Gulf of Maine species show a 40 percent variation in magnesium composition during the period of a year and indicate more rapid calcification during warmer periods (Chave, 1965).

TABLE III

SKELETAL MINERALOGY OF PRINCIPAL GROUPS OF LIVING CALCAREOUS ALGAE.

Algae	*Mineralogy*	*Environment*
RHODOPHYTA (Red)		
Corallinaceae	Calcite	Marine
Squamariaceae	Aragonite	Marine
Nemalionales	Aragonite	Marine
CHLOROPHYTA (Green)		
Codiaceae	Aragonite	Marine
Dasycladaceae	Aragonite	Marine
Charophyceae	Calcite	Nonmarine
CHRYSOPHYTA		
Coccolithophoraceae	Calcite	Marine, planktonic
CYANOPHYTA (Blue-green)	Mostly calcite	Nonmarine
PHAEOPHYTA (Brown)	Aragonite	Marine

From a survey of earlier literature, Johnson (1961) compiled a large number of chemical analyses of skeletal and nonskeletal algal material. He concluded, among other things, that the composition of marine algae varies widely with the genus, the species, and time of year.

Böhm (1973) determined the chemical composition of skeletal carbonates in several taxa of codiacean green algae. The results show variations with genera and species, and within different parts of the same plant. The mineral content of some species of *Halimeda* increases with depth.

It is well known that the three common phases of calcium carbonate have widely differing solubilities. In pure water the solubility of skeletal aragonite is about two times that of low-magnesium calcite; the solubility of skeletal high-magnesium calcite is up to 10 times that of low-magnesium calcite (Chave et al., 1952). Consequently, the original mineralogy of skeletal calcareous algae is a critical factor in their diagenesis. The selective preservation or alteration of calcareous algae in ancient limestones can be illustrated by these examples. The fine-grained aragonite skeletons of codiacean and dasycladacean green algae are invariably replaced by coarse, sparry calcite, whereas coralline red algae with calcites of varying magnesium contents show differing degrees of alteration. Coccolithophores are often preferentially preserved over other carbonate constituents because of the favorable solubility characteristics of their low-magnesium calcite skeletal elements.

Aside from paleontological considerations of fossil preservation, large accumulations of skeletal calcareous algae can be potential sources of soluble carbonate materials involved in cementation and other diagenetic processes.

CLASSIFICATION

Nowhere in paleontology is taxonomy and classification more of a problem than among calcareous algae. At once we are dealing with:

- Forms that can be classified according to biological schemes based on physiology and morphology.
- Forms presumed to be algae, but which lack living descendants or morphological counterparts.
- Biosedimentary structures resulting from the interrelationships of algae and physical processes.

In addition to these inherent difficulties, fossil calcareous algae have been treated in a casual manner by many workers for a long time -- inadequate descriptions, little attempt to follow rules of nomenclature, and no logical basis for establishing taxa, especially species, except to proliferate the literature with new names. Little wonder then that fossil calcareous algae often have been relegated to less than full stature in the systematic communities of paleontology, and have become a source of confusion to geologists attempting to use these fossils.

Living algae are classified according to pigments, nature of food reserves, chloroplast ultrastructure, and kind of flagellation. Many algae show obvious color differences -- red, green, blue-green, brown -- which are an approximate guide to a primary classification. Details of vegetative structures and reproductive processes are essential for classification into classes and smaller taxa. A comprehensive classification of the algae has been compiled recently by Levring (1969). Hommersand (1972) has summarized current information on the classification of major algal groups. The presence of skeletal carbonate per se is not a classification criterium; thus fossil calcareous algae are classified according to morphologic characteristics used in the classification schemes of analogous living forms (Table IV).

Despite an apparent understanding of the taxonomy of living algae, some botanists believe that the algae have been neglected by comparison to other plant groups, and that very little is actually known about most marine forms (Dixon, 1970). Most certainly there are problems in classifying living algae, and these uncertainties should be reflected in systematic considerations of fossil forms.

TABLE IV

CLASSIFICATION OF FOSSIL BENTHONIC SKELETAL CALCAREOUS ALGAE.

- Division: CYANOPHYTA (Blue-green algae)
  - Class: CYANOPHYCEAE
- Division: RHODOPHYTA (Red algae)
  - Class: RHODOPHYCEAE
    - Subclass: FLORIDEAE
      - Order: CRYPTONEMIALES
        - Family: SOLENOPORACEAE
        - Family: GYMNOCODIACEAE
        - Family: SQUAMARIACEAE
        - Family: CORALLINACEAE*
          - Subfamily: MELOBESIOIDEAE (Crustose corallines)
          - Subfamily: CORALLINOIDEAE (Articulated corallines)
- Division: CHLOROPHYTA (Green algae)
  - Class: CHLOROPHYCEAE
    - Order: SIPHONALES
      - Family: CODIACEAE
    - Order: DASYCLADALES
      - Family: DASYCLADACEAE
  - Class: CHAROPHYCEAE
    - Order: CHARALES
      - Family: CHARACEAE (and eight extinct families)

---

*The family Corallinaceae is shown to comprise two groups following the classifications of living marine red algae by Mason (1953) and fossil algae by Johnson (1961). However, more recently the family has been divided into six subfamilies, three each in the crustose and articulated groups (Johansen, 1969; Adey and Johansen, 1972; Adey and Macintyre, 1973).

The criteria generally used to identify and classify fossil skeletal calcareous algae are simply:

- External morphology.
- Preserved tissue, reproductive bodies, and other internal features.

The level of taxonomic discrimination depends to a very large extent on the preservation of diagnostic skeletal morphologies. Often they are absent or poorly preserved. Similarly, skeletal structures representing only a portion of the plant may be difficult to identify, whereas classification would be relatively simple if associated noncalcareous vegetative structures were also preserved.

Although external morphology is a principal characteristic for distinguishing taxa, it is known that many species of living marine algae exhibit polymorphism, that is, variations in form in response to season and different environmental conditions (Dixon, 1970). Thus in some groups each morphological variant of an alga is not a distinct taxon. Yet many genera and species of fossil algae have been based solely on minor variations in dimensions of cells, filaments, and reproductive organs. Babcock (1974) suggested that paleontologic criteria for algal classifications should be restricted to preservable diagnostic criteria used in the identification of living algae, and, in addition, should exclude dimensional variability as a basis for establishing minor taxa.

The uncertain associations between living calcareous algae and fossil forms are most pronounced in the Paleozoic where several groups of problematical algae and algae of uncertain affinities are known. Unfortunately, some paleontologists have organized meager information on some of these forms into classifications, when in fact, the amount of data is wholly inadequate to even approximate their natural relationships. The prevailing arguments about the gross biological affinities of some of these kinds of fossil groups, often whether they are plants, invertebrates, or even inorganic remains, are reasons enough for caution. Unless reasonably certain taxonomic relationships of fossils and living algae can be demonstrated, a classification serves no real purposes for understanding evolutionary patterns or for establishing the environmental ranges of fossil forms based on the habitats of analogous living species.

Summarizing, these points seem significant in considering the taxonomy and classification of skeletal calcareous algae:

- Classifications should provide a convenient method of identification and communication and also express natural relationships.
- Many fossil forms, especially in the Cenozoic and Mesozoic, can be related to major taxa of living algae; therefore, their classification should be based on preservable diagnostic criteria (hard-part morphology) used in biological schemes.
- Lacking diagnostic criteria, fossils presumed to be algae or of questionable relationships, should be classified as "problematical algae" or "algae of uncertain affinities".
- Dimensional data should be used with caution in defining minor taxa, unless the phenotypic variations of particular groups are understood.

- Generally, the identification of species should be left to the specialist.

Finally, as with any fossil group, the precision of identification required varies, and often a species or even a generic designation is unnecessary. Many paleoenvironmental analyses are possible by the recognition of larger taxa of calcareous algae.

THE FOSSIL FLORA

The true diversity of fossil skeletal calcareous algae is, at best, an estimate, because of the large number of questionable taxonomic affinities. Consequently, any compilation of described taxa requires a subjective treatment, rather than listing every taxon that has appeared in print.

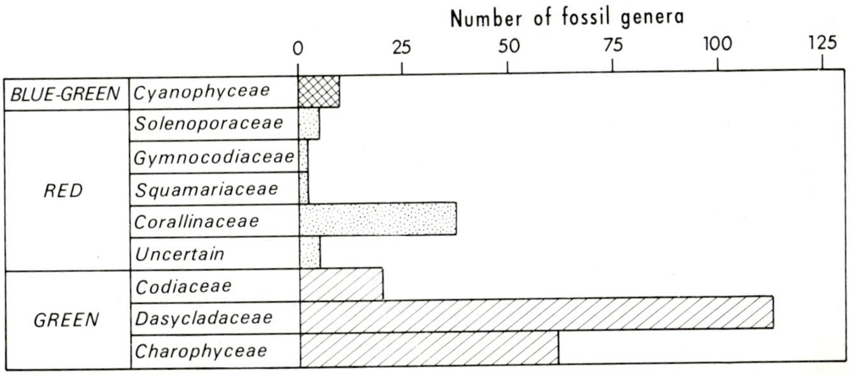

Fig. 11. Approximate number of genera of fossil skeletal calcareous algae.

An approximation of the diversity of the fossil flora is shown in Fig. 11. Some families, for example the Dasycladaceae and Corallinaceae, can be subdivided into many genera and species because of distinctive and variable structures. On the other hand, identifying autonomous members of the skeletal blue-green algae is difficult due to simple, nondescript morphologies. The relative abundance and diversity of any particular group varies with age and environment.

Chapter 3

CALCAREOUS BLUE-GREEN ALGAE (CYANOPHYTA)

Both nonskeletal and skeletal blue-green algae have been active in the deposition of carbonate sediments. Geologically, nonskeletal species have been more important because of their role in the formation of stromatolites and other algal-laminated sediments. Yet skeletal blue-green algae, though quantitatively minor, are common fossil constituents with a long geologic record.

The identity of fossil calcareous blue-green algae has been confused by the inclusion of various organisms, including foraminifera and other algal groups. Also, this group has inherited otherwise unidentifiable trace fossils and even inorganic objects presumed to belong to this phylum. Because only very simple structures characterize blue-green algae, it is difficult to prove or disprove the validity of some of these associations.

In this chapter two categories of fossil calcareous blue-green algae are considered:

- Taxa with reasonably certain affinities to living cyanophytes.
- Problematical blue-green algae.

CHARACTERISTICS

Whereas calcareous red and green algae have a variety of complex morphologies, calcareous blue-green algae are characterized by simple filaments and coccoid forms, generally of microscopic size. Most living species of blue-green algae, whether a single cell or groups of cells, are enclosed in mucilage. In filamentous forms cells occur in rows or strands, called trichomes, within a mucilaginous sheath (Fig. 12). Under certain conditions calcium carbonate is deposited in the sheath -- but not in the cells -- of some species due to algal processes (Fritch, 1945). Recently, Gleason (1972) determined the causes and nature of calcification in freshwater species in the laboratory and in natural environments in the Florida Everglades. Calcified sheaths of filamentous forms are the principal kind of skeletal carbonate produced by blue-green algae (Fig. 13).

Living blue-green algae precipitate skeletal calcium carbonate in subaerial, brackish, and freshwater environments, but not in waters of normal marine salinities. Calcareous sheaths are composed mainly of high- or low-magnesium calcites (Monty, 1967; Gleason, 1972); however, Dalrymple (1965) observed the precipitation of aragonite from supersaturated sea water within blue-green algal mats, possibly aided by bacteria. Fossil skeletal blue-green algae are widespread in marine rocks, and so there is a fundamental problem in equating this observation with their living marine counterparts which lack skeletal carbonates. Perhaps the characteristics and requirements of some ancient marine cyanophytes differed from modern forms.

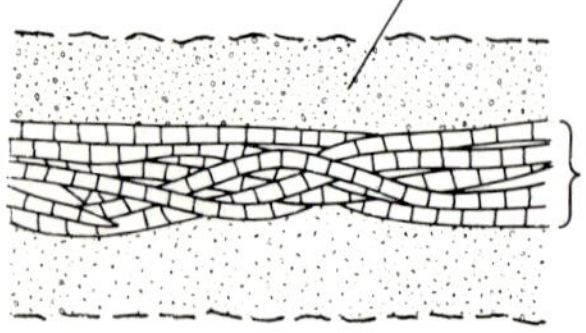

Fig. 12. Longitudinal section of filament of blue-green alga with its bundle of trichomes surrounded by a mucilaginous sheath.

Calcareous tubes of algae can result from diagenetic processes (Fig. 14). For example, Schroeder (1972) reported that the calcification of endolithic filamentous green algae in Recent reefs was due to early cementation. In the fossil record these pseudomorphs of algal filaments might be indistinguishable from calcified sheaths produced by biological processes.

CLASSIFICATION

The classification of living blue-green algae has been revised recently by Drouet (1968, 1973) who simplified the taxonomy and substantially reduced the number of taxa. The criteria used in this and other classifications of extant algae are not generally applicable to fossil calcareous blue-green algae because diagnostic nonskeletal features are not preserved. When only calcified sheaths remain, morphological variations are limited to their size, shape, and kind of branching. Drouet maintains that the blue-green algae are highly polymorphic, therefore some of these characteristics may be meaningless in distinguishing individual taxa.

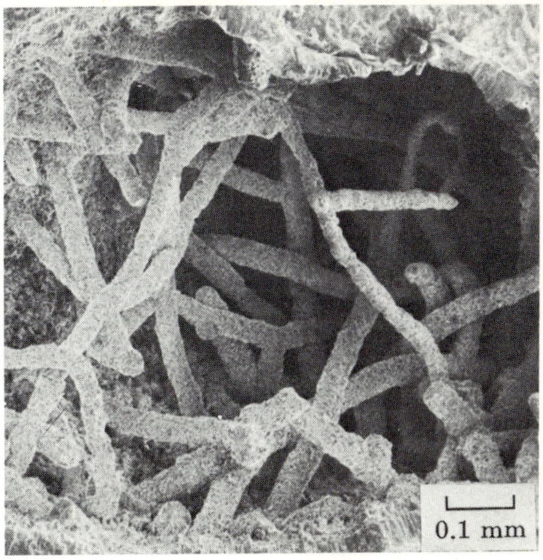

Fig. 13. Calcified sheaths of freshwater *Plectonema gloeophilum*. Recent, Aldabra Island. Scanning electron micrograph. (Courtesy of R. Riding).

Fig. 14. Diagenetically formed calcareous filaments of endolithic algae in rock cavity. Recent, Bermuda. Scanning electron micrograph. (From Schroeder, 1972).

There has been little attempt to classify fossil calcareous blue-green algae. Pia (1926) considered various filamentous forms, including *Girvanella* and *Sphaerocodium*, along with *Collenia* and *Cryptozoon*, in a discussion of blue-green algae and stromatolites. However, in 1927 Pia introduced the artificial group "Porostromata" to separate forms with a definite microstructure consisting of filaments (of unknown systematic position, but probably belonging to the green or blue-green algae) from the "Spongiostromata" (forms lacking microstructure), in which he placed algal stromatolites. Later Pia (1937) restricted the "Porostromata" to *Girvanella* and *Sphaerocodium*, and put other branched, filamentous forms, such as *Ortonella*, *Hedstroemia*, and *Bevocastria*, in the codiacean green algae. Johnson (1961) followed his classification, referring to the latter group as "crustose or nodular forms [of Codiaceae] consisting of closely packed, branching filaments".

Maslov (1956) and Maslov et al. (1963) assigned *Girvanella*, *Sphaerocodium*, *Cayeuxia*, and *Ortonella* to the calcareous blue-green algae, not recognizing the separate group of encrusting codiaceans of Pia and Johnson. Fournie (1967) compiled an illustrated bibliographic study of the Paleozoic "Porostromata" following Pia's classification.

REPRESENTATIVE GENERA

The skeletal algae considered in this section are generic names entrenched in the literature. They seem to typify the range of morphological variations observed in predominantly filamentous growth forms that most probably belong to the blue-green algae. *Ortonella* and similar branched, filamentous genera, generally classified in the Codiaceae, are included.

Girvanella

This genus is characterized by flexuous, tubular filaments of uniform diameter, composed of relatively thick, calcareous walls (Fig. 15). The tubes can best be described as unsegmented cylinders which are rarely branched. External diameters average between 10-30 microns, although specimens less than 10 microns and up to about 100 microns have been identified as *Girvanella*. Filaments may be free (loose), but usually occur in groups, twisted together to form nodules and encrusting masses on various objects (Fig. 16). The genus occurs intergrown with encrusting foraminifera (ophthalmidids) in Upper Paleozoic limestones.

Girvanella is a very common fossil with a worldwide distribution, and has been reported from the Cambrian to Cretaceous. Klement and Toomey (1967)

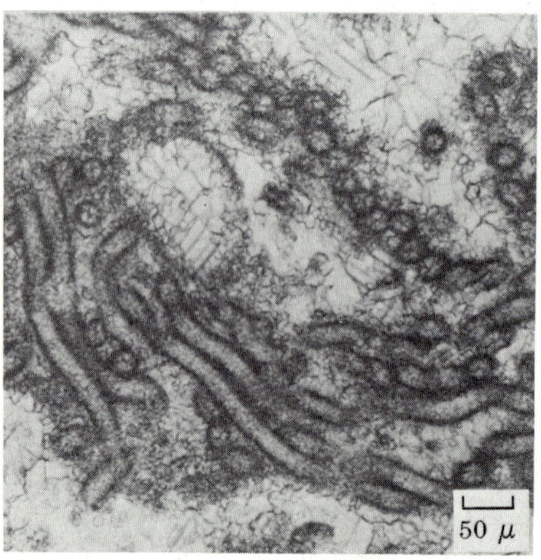

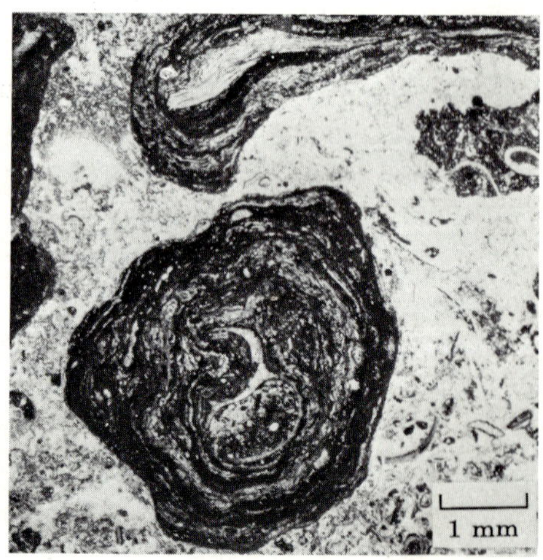

Fig. 15. *Girvanella*. Felt of interwoven tubular filaments. Devonian, Alberta. Transmitted light, thin section. (Courtesy of R. Riding).

Fig. 16. Laminated nodules and encrustations composed of *Girvanella*. Devonian, Western Australia. Transmitted light, thin section.

suggested that grain destruction in Ordovician rocks was caused by the boring and perforating action of *Girvanella*.

First described by Nicholson and Etheridge in 1878, this genus was originally regarded as a foraminifera. Bornemann (1886) concluded that *Girvanella* was a blue-green alga because of its similarity to living forms. The possible biological affinities of the genus were reviewed recently by Riding (1975). It seems likely that *Girvanella* represents the remains of numerous taxa possibly belonging to several families of blue-green algae.

Sphaerocodium

This alga consists of dichotomously branched, tubular filaments of microcrystalline calcite which developed encrusting masses. Filaments are nonseptate and branch in a distinctive fanlike pattern while maintaining contact along their inner margins (Fig. 17). This facile branching gives a

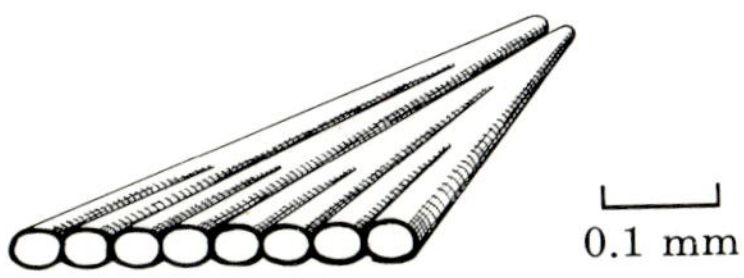

Fig. 17. *Sphaerocodium*. Sketch of branching habit of tubular filaments.

beaded appearance to groups of filaments cut transversely (Fig. 18). Filaments, commonly 30-50 microns high and 40-100 microns wide, are elliptical in cross section, being flattened parallel to the surface of attachment.

Overall growth forms are distinctively laminated and vary in shape and size, ranging from encrustations made up of only a few filament layers to masses several centimeters thick. Nodular and columnar forms are common; branching columnar masses may be several tens of centimeters high (Fig. 19). The external morphology of *Sphaerocodium* nodules and columns resembles that of algal stromatolites which formed mainly by the mechanical accretion of particles bound together by nonskeletal filamentous algae.

Sphaerocodium is primarily restricted to the Paleozoic, although Triassic forms presumably belonging to this genus are known. *Sphaerocodium* is synonomous with *Rothpletzella* and *Coactilum*, but seems to have priority over these names (Wray, 1967); however, this interpretation has been questioned by Flügel and Wolf (1969).

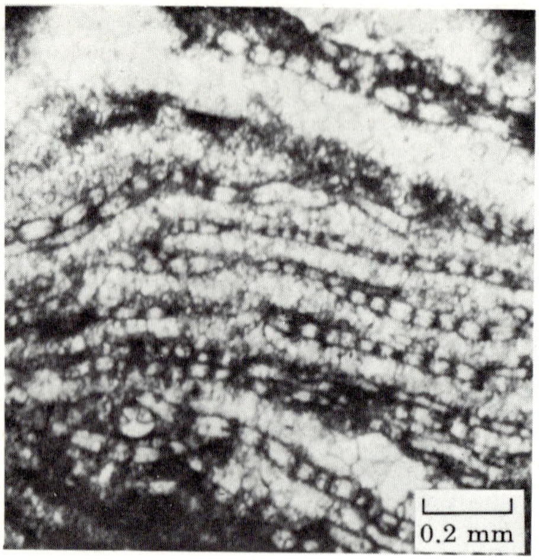

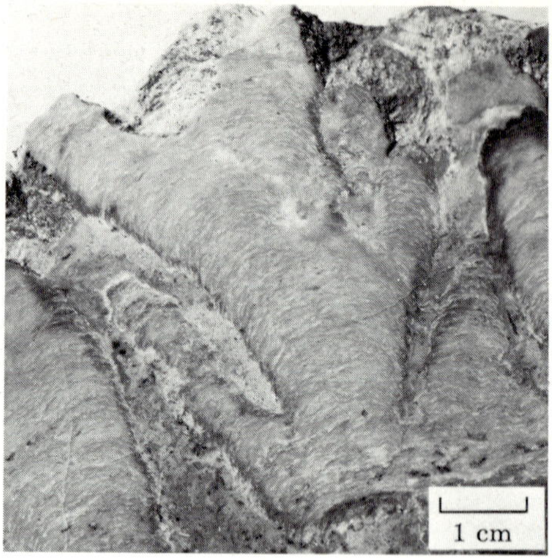

Fig. 18. *Sphaerocodium*. Transverse section of several layers of filaments. Devonian, Western Australia. Transmitted light, thin section.

Fig. 19. Columnar growth form of *Sphaerocodium*. Devonian, Western Australia. Reflected light, polished surface.

Ortonella and similar algae

Ortonella and several presumably related Paleozoic and Mesozoic genera (*Bevocastria, Cayeuxia, Garwoodia, Hedstroemia* among others) occur as nodular or crustose growth forms composed of branched filaments. Individual growth forms range from 1-10 mm in size. Filaments of *Ortonella* are circular in cross section and average 25-50 microns in diameter. They are straight or slightly undulating, and are characterized by a simple dichotomously branched habit (Fig. 20).

Genera related to *Ortonella* show different patterns of branching, but are otherwise similar morphologically (Fig. 22). *Bevocastria* has bifurcated filaments about 40 microns in diameter, but with constrictions at fairly regular intervals. *Cayeuxia* and *Garwoodia* (Fig. 23) have a right-angle branching pattern, and the filamentous branches of *Hedstroemia* occur in groups forming acute angles with each other.

The encrusting growth habit of *Ortonella* and similar forms is unlike modern codiacean green algae, which are typified by erect, segmented thalli. Thus there seems to be insufficient evidence for placing these forms in the Codiaceae (Riding, 1975), and they most probably qualify as skeletal blue-green algae. Perhaps the best grounds for this interpretation is the similiarity between *Ortonella* and the branched, calcified sheaths of the living blue-green alga *Scytonema* (Fig. 21) which was noted by Monty (1967).

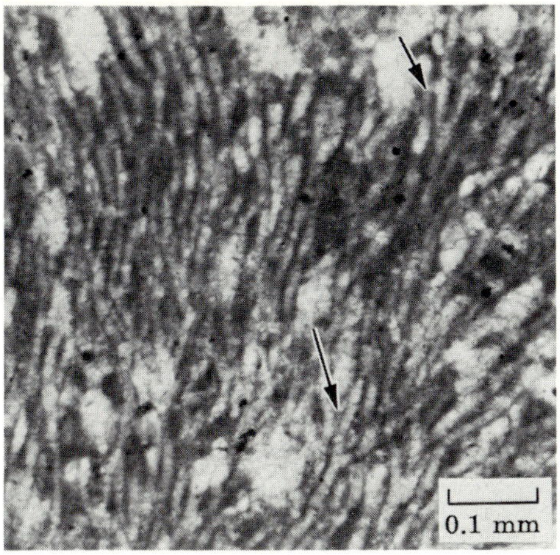

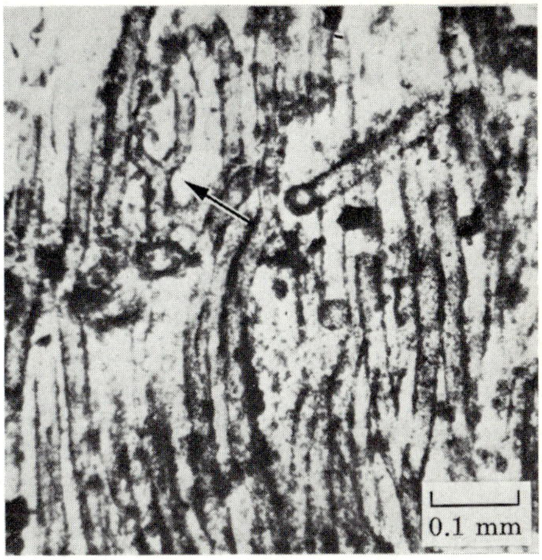

Fig. 20. *Ortonella.* Dichotomously branched filaments. Lower Carboniferous, England. Transmitted light, thin section. (Courtesy of R. Riding).

Fig. 21. Branched calcified filaments of Recent blue-green alga *Scytonema myochrous* resembling habit of *Ortonella.* Transmitted light. (From Monty, 1967).

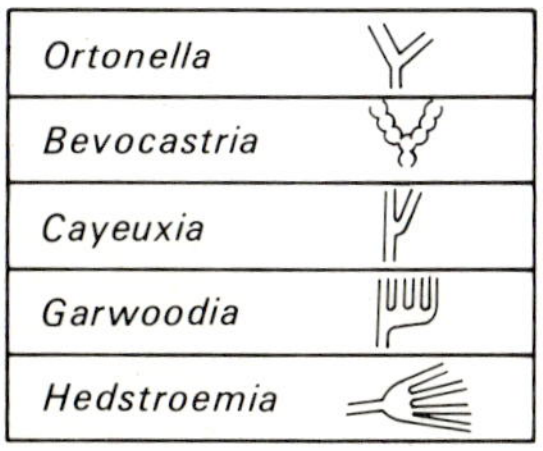

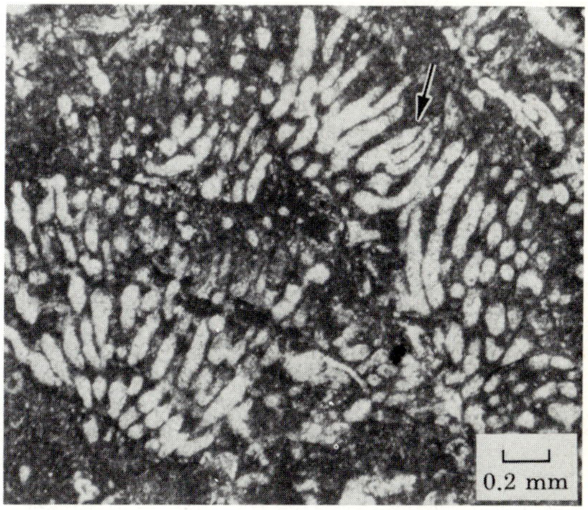

Fig. 22. Types of branched filaments in *Ortonella* and related algae. (After Johnson, 1961).

Fig. 23. *Garwoodia* showing right-angle branching pattern. Lower Carboniferous, England. Transmitted light, thin section. (Courtesy of R. Riding).

Hudson (1970) made a similar comparison of this Recent cyanophyte and Jurassic species of *Cayeuxia.*

PROBLEMATICAL BLUE-GREEN ALGAE

These taxa have been ascribed to foraminifera and various groups of algae, yet their biological affinities are far from being settled. Individual genera may be morphologically distinctive, but they lack unambiguous characteristics that can be related to living algae or other organisms.

Renalcis

Aggregates of hollow, inflated chambers characterize this genus (Fig. 24). Overall growth forms vary from simple colonies made up of a few chambers to complex botryoidal aggregates of many chambers. The size and shape of individual chambers, as well as colonies, are variable. Chambers range in size from about 50-400 microns. Colonies measure up to about 2 mm, but are often so closely packed that it is difficult to differentiate individuals. Walls are 30-80 microns thick and composed of dark microcrystalline calcite that appears opaque in thin section and porcellaneous in hand specimen.

Renalcis is similar, and presumably related, to several other genera, including *Chabakovia*, *Izhella*, and *Shuguria*. *Renalcis* occurs throughout the Cambrian, particularly in the U.S.S.R., and is also known from the Lower Ordovician and Upper Devonian. *Renalcis* has a worldwide distribution and is a quantitatively important constituent in Paleozoic reefs and other carbonate buildups.

The biological affinity of *Renalcis* (and synonymous forms) has been considered recently by Riding and Brasier (1975) who present evidence that these fossils are the earliest calcareous foraminifera. They believe it is unlikely that the chambers of *Renalcis* represent individual blue-green cells, because of their unusually large size, and conclude that its size, morphology, and composition are consistent with its being a foraminifera.

Concurrently, Hofmann (1975), supporting an algal interpretation, suggested that *Renalcis* represents the remains of irregularly arranged gelatinous colonies of nonskeletal, chroococcalean blue-green algae, in which the "walls" of the colonies have undergone early diagenesis, involving cementation. Such an interpretation, according to Hofmann, would account for the great variation in size, implying that the "chambers" are not individual cells but colonies of cells.

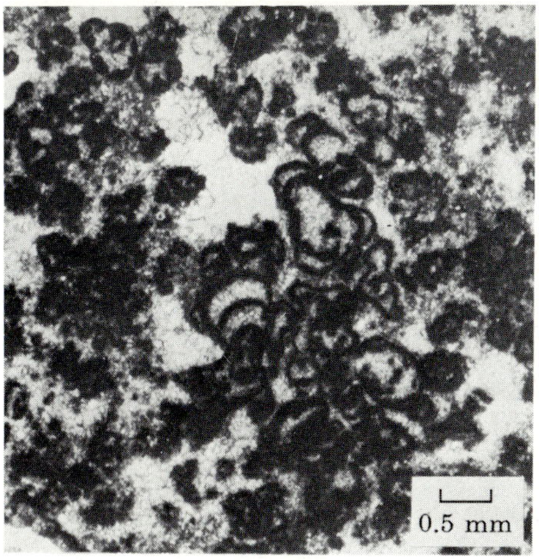

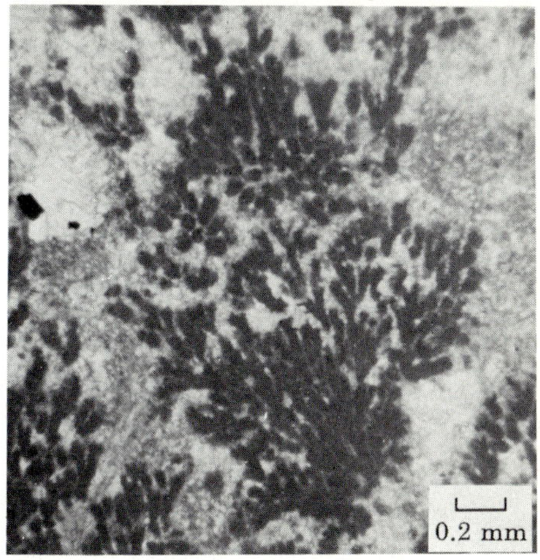

Fig. 24. *Renalcis*. Characteristic growth form of inflated chambers. Devonian, Western Australia. Transmitted light, thin section.

Fig. 25. *Epiphyton*. Cluster of densely branched "filaments". Cambrian, Virginia. Transmitted light, thin section. (Courtesy of J. F. Read).

Epiphyton

Epiphyton consists of radiating clusters of densely branched thalli, and resembles a small bushy plant, generally a millimeter or less in overall dimension (Fig. 25). Individual stems (?filaments) branch dichotomously; they are circular in cross section, solid, unsegmented, and rarely show microstructure.

This dendritic fossil is common throughout the Cambrian, and similar forms have been reported from the Devonian (Wray, 1967). *Epiphyton* is a widespread constituent in Lower Paleozoic carbonate accumulations.

Epiphyton has been regarded variously as a blue-green, green, and red alga (Riding and Wray, 1972). According to Korde (1959), specimens from the Cambrian of Siberia have a cellular microstructure together with what she regarded as cell-wall pores and sporangia; consequently, she classified this alga in the Rhodophyta. Subsequent comprehensive studies of Cambrian *Epiphyton* by Korde (1961, 1973) have resulted in the description of numerous genera and species that have been organized into an elaborate classification of fossil red algae.

Unequivocal cellular tissue in *Epiphyton* would satisfy some arguments for classifying this genus in the red algae. However, most specimens do not possess any microstructure, and published illustrations of these features are not definitive. In many respects *Epiphyton* resembles *Renalcis* and *Frutexites*, and on this basis it is considered a problematical blue-green alga.

Frutexites

This genus has the appearance of minute "shrubs" less than 1 mm high, generally disposed vertically on bedding planes and arranged in groups (Fig. 26). They consist of incompletely branched "filaments" about 50 microns in diameter.

Maslov (1960) suggested that this genus belongs to the blue-green algae. However, he pointed out that the fossil did not preserve the original cells, but rather corresponds to the external form of a mucilaginous casing enveloping groups of cells -- a similar interpretation to the one proposed by Hofmann (1975) for *Renalcis*. *Frutexites* was originally described from Ordovician stromatolites of the U.S.S.R. It has subsequently been reported in Jurassic stromatolites of Poland (Szulcsewski, 1963), and in Devonian stromatolites of Western Australia (Playford et al., 1976). In each case *Frutexites* is characterized by the concentration of iron oxide in its filaments.

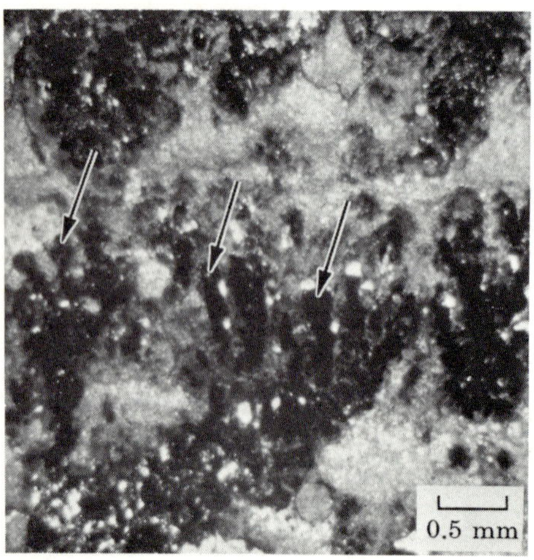

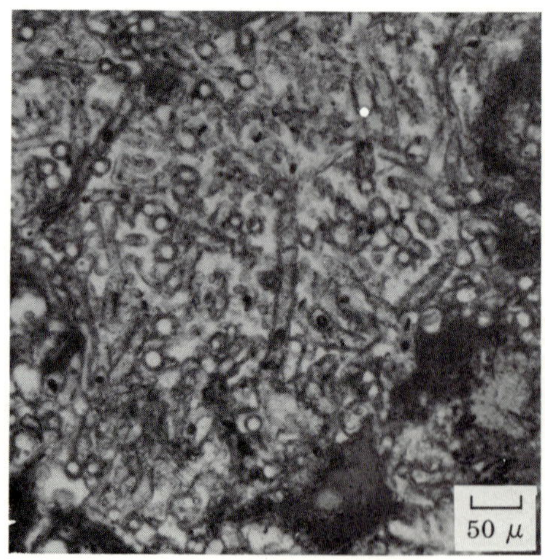

Fig. 26. *Frutexites*. Vertical section of numerous layers of individual "shrubs". Devonian, Western Australia. Transmitted light, thin section.

Fig. 27. Network of calcareous tubules, possibly blue-green algae, in Recent caliche deposits, Barbados. Transmitted light, thin section. (James, 1972).

GEOLOGIC RANGE

Most fossil skeletal blue-green algae occur in Paleozoic rocks rather than younger sediments, and they seem to be unrecorded in the Cenozoic (Fig. 28). The first occurrence of a high proportion of these algae is in early Paleozoic time. The overall distribution pattern is puzzling because if these algae do represent the calcified sheaths of several kinds of filamentous blue-green algae then they should occur throughout the Phanerozoic.

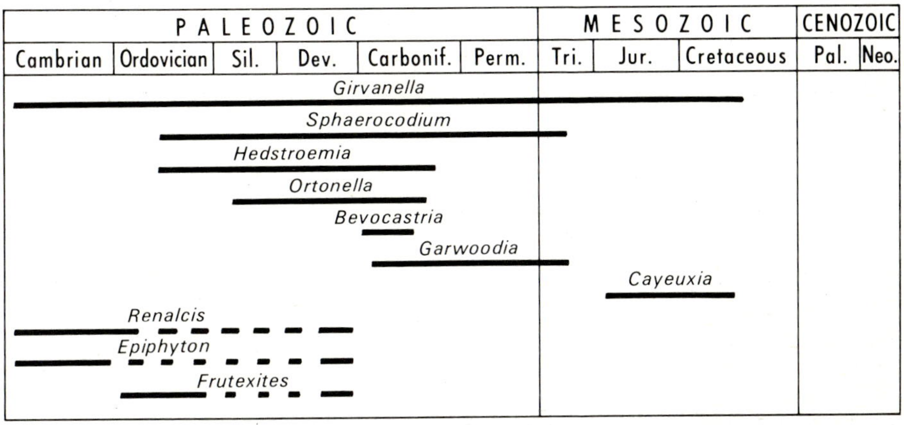

Fig. 28. Geologic ranges of common genera of calcareous blue-green algae.

ENVIRONMENTAL DISTRIBUTION

Living blue-green algae inhabit a wide range of marine and nonmarine environments, but apparently no living species secrete calcium carbonate in normal marine waters. Nevertheless, several fossil skeletal genera with filamentous habits are widespread in marine rocks of various ages. Consequently, it would seem that the observed occurrences of fossil blue-green algae are the best guide to their environmental distribution in the ancient record, rather than by analogy with living forms.

Riding (1975) discussed the problems of using *Girvanella* and other algae as depth indicators, citing the case that this genus, along with some other skeletal blue-green algae, represent the remains of several taxa. As such it seems impossible to set limits on their depth of occurrence.

Girvanella has been reported most often from marine environments, generally in shallow shelf, carbonate facies (less than 50 m), although it has been recorded in nonmarine limestones (Berryhill et al., 1971). Also, calcified filaments preserved in subaerially formed Recent caliche in Barbados (Fig. 27) could be identified as *Girvanella* (James, 1972). Thus, this kind of alga seemed to occur in a variety of nonmarine and shallow marine environments, and by itself could not be used to discriminate marine from nonmarine facies. *Ortonella* and related forms with branched filaments have a distribution similar to *Girvanella*.

Other calcareous blue-green algae, such as *Sphaerocodium* and *Renalcis*, are restricted to marine sediments, and a species of *Sphaerocodium* in the Devonian reef complexes of Western Australia occurs in deep water, fore-reef facies (Wray, 1972).

Chapter 4

CALCAREOUS RED ALGAE (RHODOPHYTA)

The red algae are a large group of dominantly benthonic, marine plants; only a relatively few species (about 2 percent) live in fresh water. Most forms are characterized by rather complex thalli which are distinctively colored by red pigments. Biologically, the group shows considerable uniformity and may represent a long evolutionary history, possibly having evolved from the blue-green algae during a very early time. The Rhodophyta seem to be intermediate between the Cyanophyta and Chlorophyta, having plastid characteristics similar to the blue-green algae, and a eukaryotic organization, a nuclear division (mitosis), and wall characters similar to the green algae.

Calcareous red algae may be assigned to four families:

- Solenoporaceae
- Gymnocodiaceae
- Squamariaceae
- Corallinaceae

The first two listed are extinct. The latter two families are represented by both living and extinct members. The Corallinaceae are a group of thoroughly calcified plants with a record extending back to mid-Mesozoic time. In addition, fossil red algae include a number of taxa, mainly Paleozoic forms, that do not fit into these four families. They have been variously classified formally in other families, and informally assigned to "problematical red algae", "red algae of uncertain affinities", and "ancestral corallines".

Korde (1973) has presented a detailed classification of Cambrian red algae. The scheme proposed two new classes subdivided into six new orders and 17 new families, plus two previously recognized families, including the Solenoporaceae. The validity of this classification is unclear, because many of the taxa are new and known only from the U.S.S.R., while the true affinities of others are still uncertain.

SOLENOPORACEAE

Characteristics

This extinct family is characterized by encrusting forms that developed nodular growth habits ranging in size from several millimeters to a few centimeters (Fig. 29). Rounded, hemispherical masses seem to be the most common shapes, although thin crusts are known.

Internally, these algae consist of radially or vertically divergent calcified rows or filaments of cells which are circular or polygonal in cross sections (Figs. 30, 31). Cell diameters range from 20-100 microns, averaging larger than those of most modern coralline algae. Cellular tissue is relatively simple compared to the corallines, and is undifferentiated, that is, only one kind of cellular arrangement occurs in the same genus. The nature of partitions between cells within filaments is an important generic character. They may be virtually absent, well developed, and regular or irregular in their spacing. Johnson (1960) believed the original plant tissue in all Solenoporaceae had horizontally partitioned filaments, but the amount of calcification was variable; thus some horizontal cell walls were preserved and others were not.

Definitive reproductive organs in the Solenoporaceae are usually absent, and many structures that have been interpreted as having this function are doubtful and obscure. Pia (1927) assumed that these organs were external and not calcified, but recently, Elliott (1973) described a Miocene genus (*Neosolenopora*) with calcified reproductive structures -- a development at the very end of solenoporacean evolution.

Classification

Solenopora and related forms have occasionally been referred to animal groups, but usually they are considered calcareous algae and assigned to the Rhodophyta, presumably related to living Corallinaceae. Earlier, questions were raised as to whether this group should be an independent family or subfamily rank of the Corallinaceae. Now there is general agreement that the Solenoporaceae constitute a coherent group of extinct algae separate from the Corallinaceae, although solenoporaceans probably provided an ancestral stock to the true corallines.

Solenoporacean genera are classified almost entirely on the type of cellular tissue. Aspects considered important are:

- Presence or absence of filament partitions (horizontal cell walls).
- Regular or irregular spacing of partitions.
- Shape of cells in transverse section.

Johnson (1960) reviewed the characteristics and classification of Paleozoic Solenoporaceae. Numerous species of solenoporacean genera have been distinguished on the basis of cell dimensions alone; for example, nearly 40 Paleozoic species of *Solenopora* have been described. This practice seems suspect based on our knowledge of living red algae which show dimensional variations of tissue within the same species.

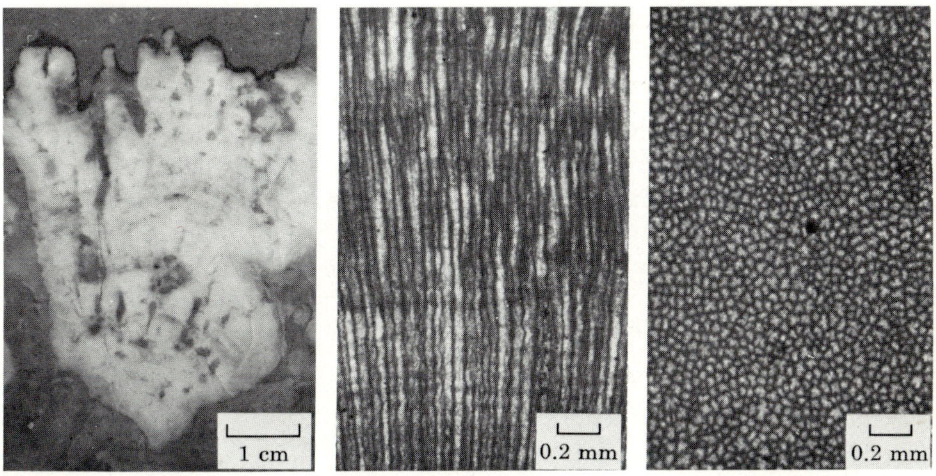

Fig. 29. *Parachaetetes*. Vertical section of thallus. Devonian, Alberta. Reflected light, polished surface.

Fig. 30. *Solenopora*. Vertical section of cell filaments. Ordovician, Kentucky. Transmitted light, thin section.

Fig. 31. *Solenopora*. Transverse section of cells. Ordovician, Kentucky. Transmitted light, thin section.

Representative genera

About 10-12 genera have been assigned to the Solenoporaceae, although only a few of these occur commonly or seem to be valid taxa. Some appear to be synonymous with *Solenopora* or *Parachaetetes* (e.g., *Pseudochaetetes*), and *Pycnoporidium* probably belongs to the green algae (Elliott, 1963).

Solenopora

Rounded, nodular masses are the characteristic growth form of this genus. In vertical section, the most apparent features are vertical or radiating walls between filaments (Figs. 30, 31). Partitions between cells within filaments are absent or inconspicuous. In transverse section, cells are rounded to polygonal, averaging 30-50 microns in diameter. Reproductive organs are not known; presumably they were external and uncalcified.

Parachaetetes

The thallus of *Parachaetetes* is the same general shape as *Solenopora*, namely nodular masses which are often hemispherical. Cellular tissue is compact and composed of radially arranged, rounded or polygonal filaments of cells about the same size as those in *Solenopora*. In contrast to *Solenopora*, filaments have well defined, regularly spaced partitions between cells, which gives the tissue a gridlike pattern in vertical sections (Fig. 32). Length of individual cells is generally greater than the cell diameter. Some forms show distinctly banded growth layers in vertical section. Reproductive organs were not preserved.

Solenomeris

This Cenozoic genus is characterized by a marked irregularity of its cellular tissue (Fig. 33). Cells are polygonal in transverse section, averaging 40-60 microns in diameter. In vertical section, individual cells appear to be irregular in shape, because cell partitions alternate in position with adjacent filaments to give a zigzag effect. The thallus varies in shape from nodular forms to thin encrusting species. Reproductive organs were probably external and uncalcified, although internal conceptacles have been described.

Geologic range

The Solenoporaceae are predominantly Paleozoic and Mesozoic algae, yet a few taxa range well into the Cenozoic (Fig. 34). Two genera, *Solenopora* and *Parachaetetes*, have exceedingly long geologic ranges, but with some gaps in their record. *Parachaetetes* has only recently been reported from the Late Carboniferous (Pennsylvanian), where previously solenoporaceans were unknown (Heckel, 1975). Members of this family seem to be most common and widespread

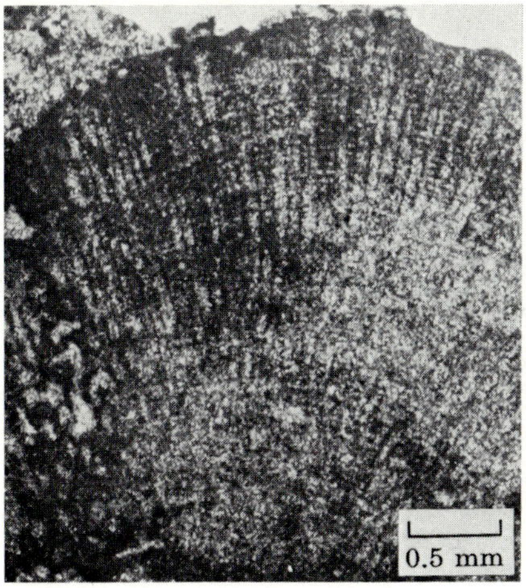

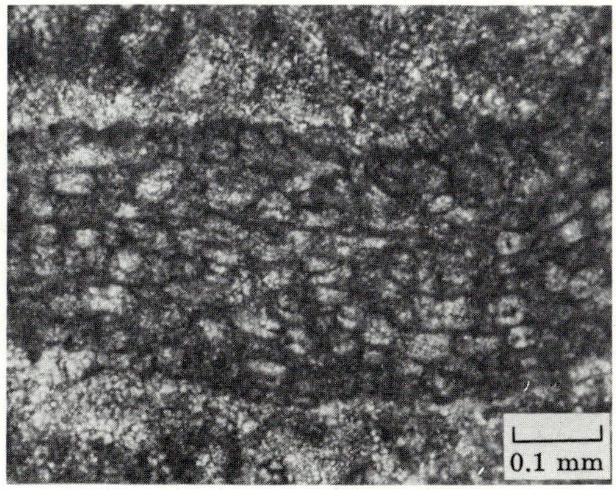

Fig. 32. *Parachaetetes*. Vertical section. Devonian, Western Australia. Transmitted light, thin section.

Fig. 33. *Solenomeris*. Paleocene, Libya. Transmitted light, thin section.

in the Paleozoic, with subordinate numbers in the Mesozoic. The few genera and species that survive into the early part of the Cenozoic can be locally abundant.

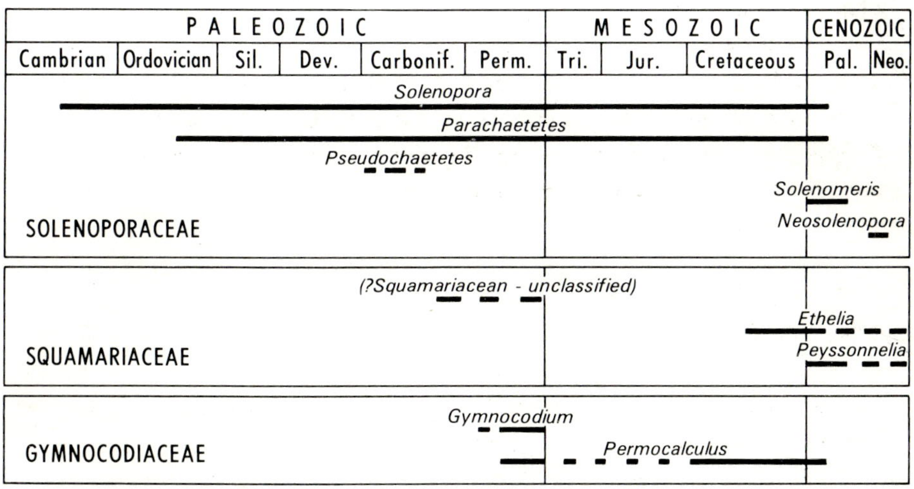

Fig. 34. Geologic ranges of solenoporacean, squamariacean, and gymnocodiacean genera.

Environmental distribution

The environmental distribution of the Solenoporaceae is comparable to some modern coralline algae, but this extinct group was not as cosmopolitan in its depth and temperature ranges as the Corallinaceae. The sedimentological record indicates that the Solenoporaceae occupied open-marine environments of normal salinities. These fossils are associated with a variety of marine invertebrate constituents in Paleozoic rocks and occur most often in shelf carbonates, including bioherms and reef deposits.

Elliott (1965a) noted that Mesozoic solenoporaceans did not occupy the same ecological niche with reef-building hexacorals, because the two usually occur in the different facies. However, Cenozoic solenoporaceans seem to have adapted to a wider range of environments because they occur with hexacorals and coralline algae. Solenoporaceans are a conspicuous but minor element in shelf and reef facies in Paleocene limestones of Libya (Wray, 1969).

GYMNOCODIACEAE

Characteristics and classification

Calcareous algae assigned to this extinct family have a contrasting morphology to the compact, cellular tissue characterizing other fossil red algae. Gymnocodiaceans are believed to be the remains of erect, branched plants similar to the living marine red alga *Galaxaura*, belonging to the family Chaetangiaceae (order Nemalionales).

Only two genera, *Gymnocodium* and *Permocalculus*, are assigned to the Gymnocodiaceae, and each is known only from perforate, calcareous segments and fragments (Figs. 35, 36). Preserved thalli are elongate cylinders, often showing alternate pinching and swelling. The two genera are similar to each other, but *Permocalculus* is distinguished by smaller and more numerous pores.

The genus *Gymnocodium* was named by Pia, who considered the fossil a dasycladacean, but later he transferred it to the Codiaceae. In 1937 Pia assigned the genus to the Chaetangiaceae and compared it to the Recent genus *Galaxaura*. Elliott (1955) supported this interpretation and erected a new family (Gymnocodiaceae) to accommodate these extinct algae.

Recent species of *Galaxaura* usually have minor amounts of calcification not comparable to the apparent extent of skeletal material preserved in fossil gymnocodiaceans, and the classification of these fossils has been questioned on other grounds. However, Elliott (1961) recognized new morphological features in *Permocalculus* that he believed analogous to characteristics in *Galaxaura*, thus reinforcing the supposed relationship between the fossil and living forms.

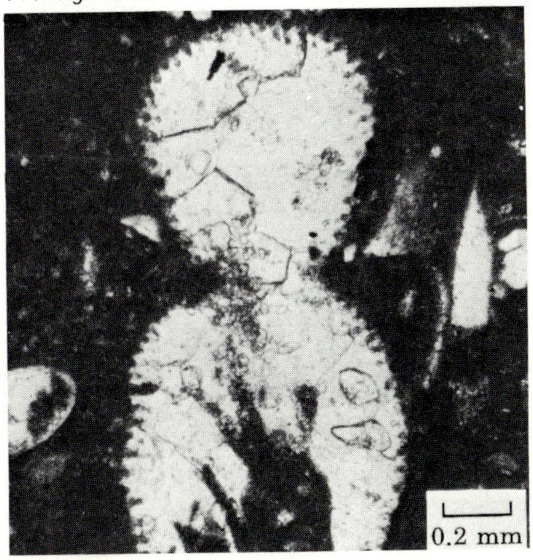

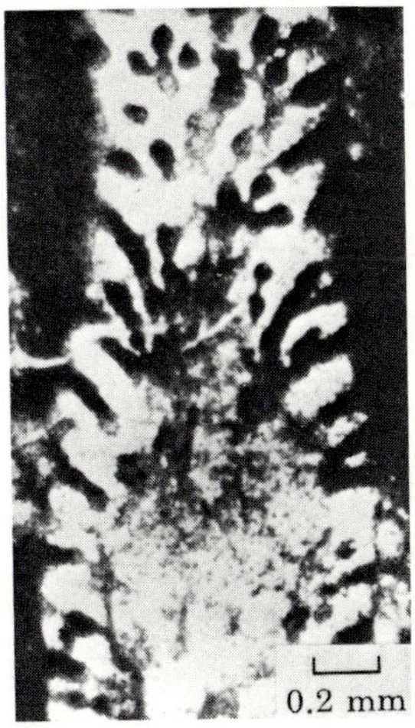

Fig. 35. *Permocalculus*. Cretaceous, Texas. Transmitted light, thin section.

Fig. 36. *Gymnocodium*. Permian, Austria. Transmitted light, thin section. (From Pia, 1937).

Geologic range and environmental distribution

Gymnocodiaceans have limited and discontinuous geologic ranges (Fig. 34). *Gymnocodium* is known only from the Permian, whereas *Permocalculus* occurs in Permian, Cretaceous, and Paleocene rocks. Both genera are geographically widespread and have been described from numerous localities in Asia, Europe, and North America. These fossil algae are often associated with codiaceans and dasycladaceans, and presumably occupied similar shelf environments (Elliott, 1958).

SQUAMARIACEAE

Characteristics

The Squamariaceae are encrusting plants composed of cellular tissue. Many living species grow as flattened crusts firmly cemented to a solid substrate, while others are petaloid forms exposed on all surfaces except for a basal attachment. Crusts are 0.2-0.5 mm thick and extend over areas ranging from a few millimeters to at least 10 cm. Thalli vary from simple, flattened crusts to complex undulating, foliate forms (Fig. 40).

A few Recent genera have thoroughly calcified thalli of aragonite, while others have little or no calcification. Two living calcareous genera with widespread distribution, *Peyssonnelia* and *Ethelia*, are known from the fossil record. The growth form and internal structure of calcareous squamariaceans resemble the morphology of some crustose Corallinaceae.

Internally, thalli consist of closely packed, cellular filaments differentiated into a basal layer (hypothallium), from which arise vertical or arcuate rows of cells (perithallium) (Fig. 37). Some forms have a medullary hypothallium with opposing layers of perithallial tissue. Hypothallial cells are more or less rectangular, measuring about 30 microns high and 20 microns

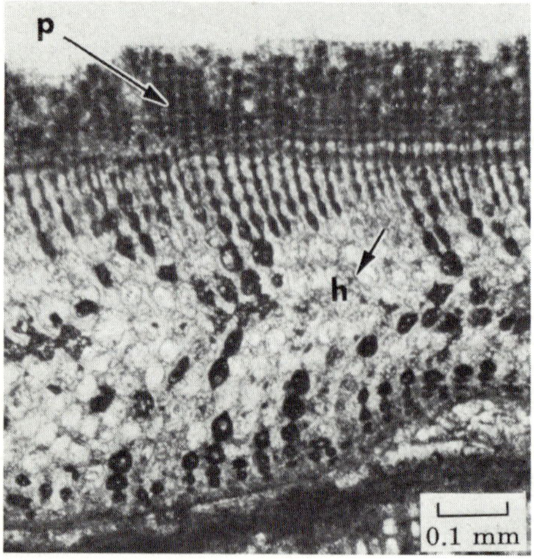

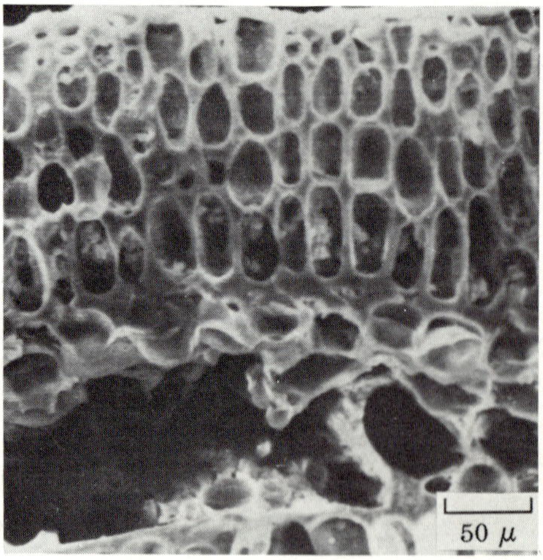

Fig. 37. *Ethelia* (=*Polystrata*). Vertical section of thallus; hypothallium (h), perithallium (p). Recent, Hawaiian Islands. Transmitted light, thin section.

Fig. 38. Squamariacean. Details of perithallium. Recent, Caribbean. Scanning electron micrograph.

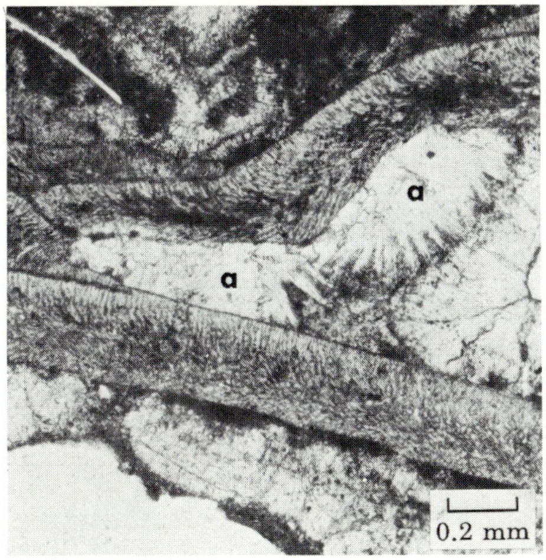

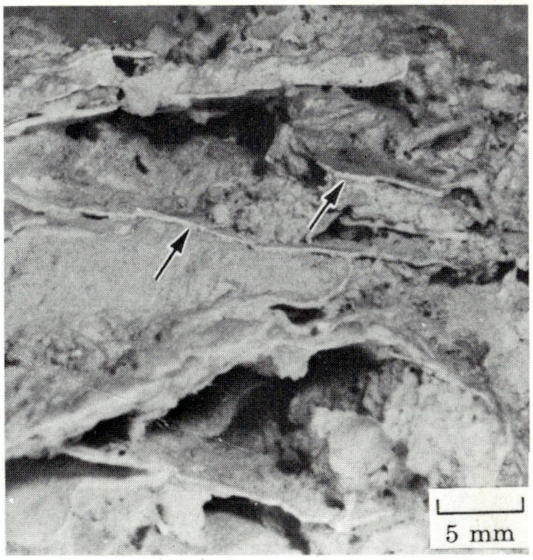

Fig. 39. Squamariacean. Sections of thalli and associated radial fans of aragonite (a). Recent, Belize. Transmitted light, thin section.

Fig. 40. Squamariacean. Complex of thalli in lithified sediment. Recent, Belize. Reflected light. (Courtesy of R. N. Ginsburg).

wide, and perithallial cells are somewhat smaller (Fig. 38). There is, however, a considerable variation in cell dimensions. Reproductive organs are borne in wart-like protuberances on the outer surface of living species, but these have not been observed in fossils.

A yellow-brown tint is observed in thin sections of fossil squamariaceans, in contrast to the dark gray cast and fine-grained texture of the Corallinaceae. No doubt this is related to the altered aragonitic mineralogy of squamariaceans compared to the calcitic composition of the corallines. Attached radial fans of aragonite, interpreted as early diagenetic products (Ginsburg and James, 1976), have been found on Recent squamariaceans in lithified sediments (Fig. 39). Similar fabrics have been observed in association with unidentifiable Paleozoic phylloid algae (Wray et al., 1975), suggesting a possible genetic relationship.

Geologic range and environmental distribution

The geologic record of squamariaceans is quite incomplete (Fig. 34). The extant genera, *Ethelia* and *Peyssonnelia*, are known as fossils from various localities, especially in Lower Cenozoic limestones (Johnson, 1964a, 1965; Massieux and Denizot, 1964; Denizot and Massieux, 1965). Phylloid algae

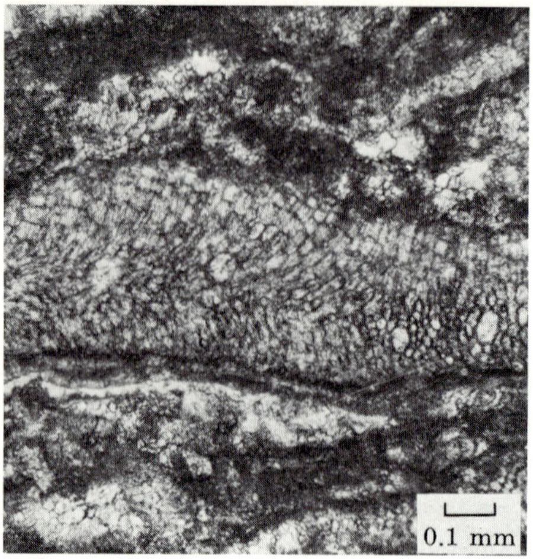

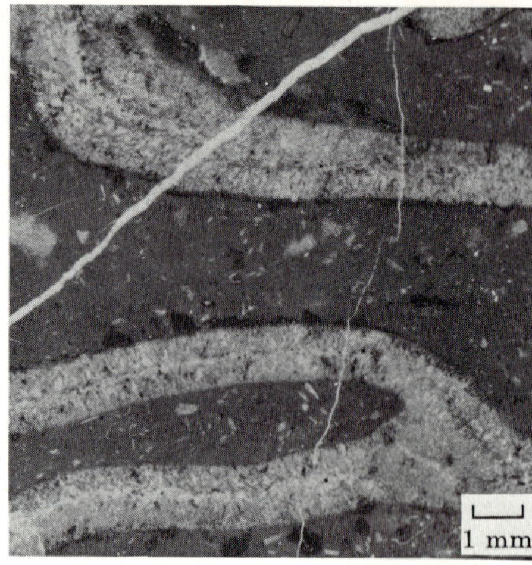

Fig. 41. *Ethelia*. Paleocene, Libya. Transmitted light, thin section.

Fig. 42. ?Squamariacean. Permian, Tunisia. Transmitted light, thin section.

suspected of belonging to the Squamariaceae occur in the Late Carboniferous (Pennsylvanian) and Permian (Fig. 42); however, the internal structure is not definitive, thus the interpretation is tentative.

Living calcareous Squamariaceae occur in tropical and subtropical marine environments, although uncalcified forms extend to subarctic regions. They are found abundantly in normal marine salinities at shallow depths from just below low tide to a few meters. Species also extend to depths of 30-50 m in Caribbean reefs and have been recovered from depths of 90 m in the Hawaiian Islands.

Most forms require a firm substrate for attachment, but unattached species of *Peyssonnelia* living on soft bottoms occur in the Mediterranean. Fossil forms are associated with scleractinian corals and coralline algae in Paleocene reef facies of Libya (Fig. 41), and also occur in shelf carbonates of comparable ages in the Middle East and Guatemala.

CORALLINACEAE

The Corallinaceae is the largest family of the order Cryptonemiales and includes a great diversity of forms, almost all of which are thoroughly calcified. For these reasons, the corallines have a rather complete fossil record consisting of numerous genera and species. These algae are some of the more common and widespread skeletal constituents in Cretaceous and Cenozoic marine carbonate facies.

Characteristics

Calcified cellular tissue is the most striking feature of these algae and individual genera are characterized by definite arrangements and sizes of cells. Reproductive organs are internal and preserved within the cellular tissue. In some Corallinaceae these organs occur in conceptacles, which open to the exterior by one or more pores, while in others individual sporangia occur in rows and are not in conceptacles. The Corallinaceae have two distinctly different growth habits, one group comprising crusts, nodules, and rigid branched forms (subfamily Melobesioideae), and the other consisting of erect, articulated or jointed corallines (subfamily Corallinoideae).

Growth form

Members of the Corallinaceae have exceedingly variable growth habits. Crustose corallines are generally attached, but some are unattached or free. Encrusting genera vary from thin (a single-cell-layer thick) crusts about 1 mm across to massive and nodular developments 10 cm or more in diameter. This group also includes forms consisting of multiple crusts and rigid branched habits (Fig. 1).

The articulated corallines are rather uniform in shape. These plants are attached and have erect, jointed, branched thalli (Fig. 43). Calcified segments (intergenicula) are connected together by uncalcified joints or

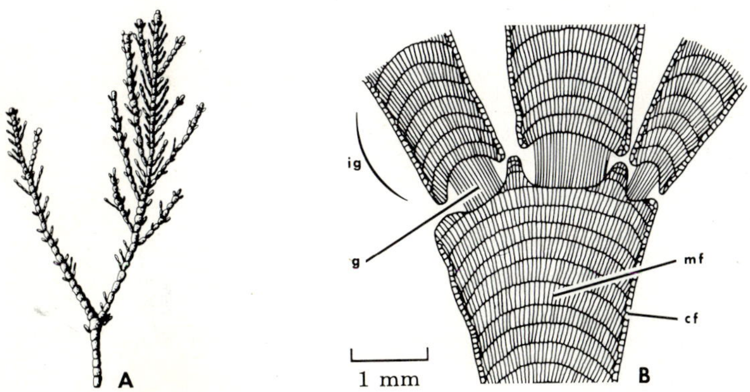

Fig. 43. Articulated coralline alga. A) Typical growth form (*Corallina*). B) Internal structure of calcareous segments. Geniculum (g); intergeniculum (ig); medullary filaments (mf); cortical filaments (cf).

nodes (genicula). Individual segments vary in shape from cylindrical to flattened forms. These algae disarticulate upon death, consequently fossils usually occur as separate segments.

Cellular tissue

The basic anatomical structure of the Corallinaceae consists of closely packed, partitioned filaments, the cell walls of which are heavily calcified. Cellular tissue of most crustose coralline genera is differentiated into three distinct regions: hypothallium, perithallium, and epithallium.

Hypothallial filaments are oriented more or less parallel to the substrate, forming the basal-most tissue (Fig. 44). Basal hypothallium may develop as a multilayered type consisting of many cell layers or as a single-layered type, which may have a single layer of cells elongate parallel to the substrate, equidimensional, or elongate perpendicular to the substrate (called palisade). Multilayered hypothallial tissue also occurs in arcuate layers, called coaxial.

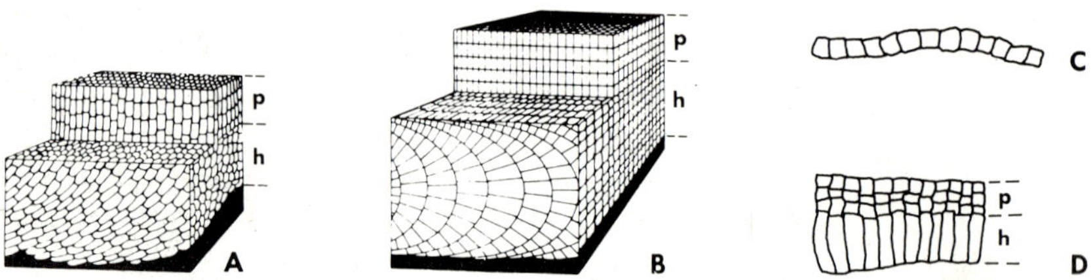

Fig. 44. Types of hypothallial tissue in crustose corallines. A) Multilayered. B) Multilayered coaxial. C) Single-layered equidimensional. D) Single-layered palisade. Perithallium (p); hypothallium (h).

The perithallium develops above or upon the hypothallium and consists of filaments oriented perpendicular to the substrate. The thickness of perithallial tissue is highly variable. In some genera it is virtually absent or very thin, but in others it forms a large part of the plant tissue. Cell dimensions of the perithallium are generally less than those of the hypothallium.

Perithallial tissue of a few crustose coralline genera is characterized by occasional larger cells than those in the surrounding tissue. Called heterocysts, these enlarged cells may occur singly, as a vertical heterocyst chain, or in horizontal rows (Figs. 57, 58).

The epithallium is a thin surface layer of uncalcified cells, often called cover cells. This outermost layer of cells is separated from the perithallium by a region of cell layers, the meristem, in which most of the vegetative cell division occurs. The epithallium and meristem region are not preserved in fossil corallines, simply because they are not calcified.

Cells of the same filament in species of the subclass Florideae are connected by minute features called primary pits and these are especially apparent in the calcified thalli of the Corallinaceae (Fig. 45). These structures are passages through the cell wall which are partially blocked by plates or plugs. In addition, some coralline genera are characterized by micron-sized secondary-pit connections between cells of adjacent filaments.

The terminology applied to crustose corallines is not strictly applicable to the articulated corallines. Rather, calcified cellular organization within an individual segment is differentiated into medullary tissue and cortical tissue (Fig. 43). Segments consist of a central core of partitioned filaments, the medula, which extend upward to the next segment. The medula is surrounded by cellular filaments (cortex) which develop from it.

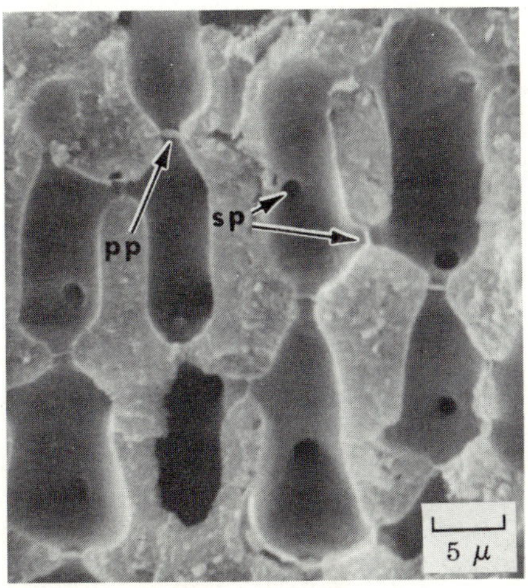

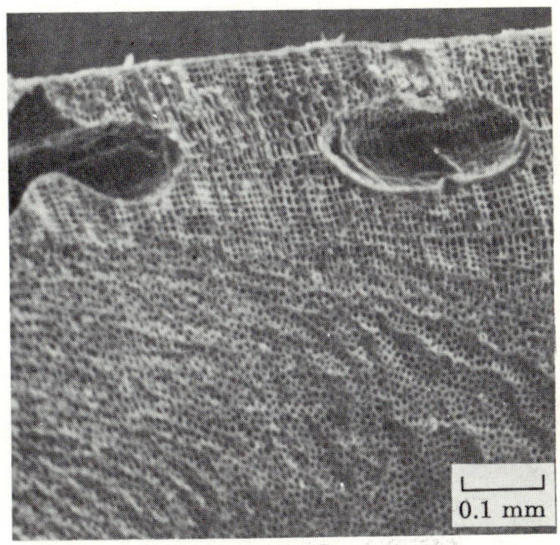

Fig. 45. *Lithophyllum*. Detail of perithallium showing primary-pit connections (pp) between cells within same filament and secondary pits (sp) connecting adjacent filaments. Recent, Guam. Scanning electron micrograph.

Fig. 46. *Lithophyllum*. Vertical section of perithallium showing single-pored conceptacles. Top of specimen is growth surface. Recent, California. Scanning electron micrograph.

Reproductive structures

Reproductive organs are characteristic morphological features of the Corallinaceae. In crustose corallines these generally develop on the outer surface of the thallus and are preserved within perithallial tissue (Fig. 47).

Reproductive bodies may occur as individual sporangia in rows (and are said to form in sori), or as multiple sporangia within conceptacles. Conceptacles are of two types, those with a single opening or pore (Fig. 46), and multi-pored structures.

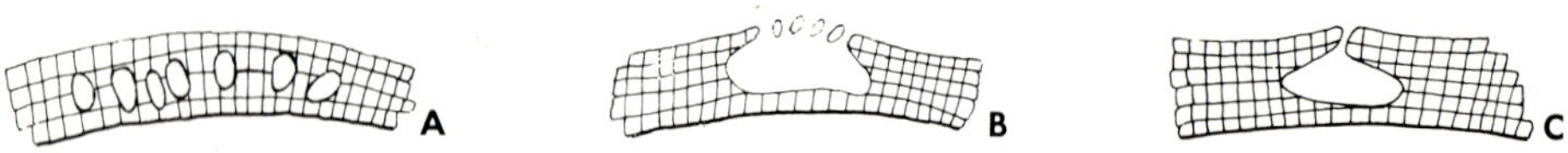

Fig. 47. Types of reproductive organs in crustose coralline algae. A) Sporangial sori in *Archaeolithothamnium*. B) Multipored conceptacle. C) Single-pored conceptacle.

Articulated corallines have single-pored conceptacles only, which originate in the medullary and cortical tissues. Those developing in medullary tissue are designated axial or marginal conceptacles; lateral conceptacles form in cortical tissue.

Classification

Preserved skeletal features of the Corallinaceae provide a basis for identifying many fossil genera, yet some living genera cannot be recognized in the geologic record because they are characterized on the basis of non-preservable structures. The generic-level taxonomy of the common coralline genera has changed little since it was established, and the basic elements have been summarized by Kylin (1956) and Johnson (1961). However, more recent detailed morphological studies have provided new information on inter-generic relationships and resulted in revised classification schemes (Johansen, 1969; Adey and Johansen, 1972; Cabioch, 1972; Littler, 1972; Adey and Macintyre, 1973).

Adey and Johansen (1972) distinguished six subfamilies in the Corallinaceae (three each in the crustose and articulated groups) on the presence or absence of genicula, of tetrasporangial plugs (i.e., number of pores in conceptacles), and secondary-pit connections. Although the first two criteria may be recognized in fossils, secondary pits are impossible to discern in thin sections of fossil corallines.

Diagnostic generic-level criteria used to identify fossil corallines include:

- Type and location of conceptacles.
- Character of hypothallium.

- Character of perithallium.
- Presence or absence of heterocysts and their character.

In addition to these criteria, nonpreservable features, such as meristems and epithallia, distinguish certain living crustose corallines. The classification of Adey and Johansen has helped clarify generic concepts of the Corallinaceae, and this approach has been used to gain new insight into possible phylogenetic trends in the crustose corallines (Fig. 68); however, the introduction of new suprageneric taxa seems an unnecessary complication in dealing with fossil assemblages.

Fossil Corallinaceae are best handled by classifying them into the established subfamilies Melobesioideae (crustose corallines) and Corallinoideae (articulated corallines). Preservable morphological features of the cellular tissue and reproductive organs of the crustose coralline algae can be combined into a relatively simple characterization of common genera (Fig. 48).

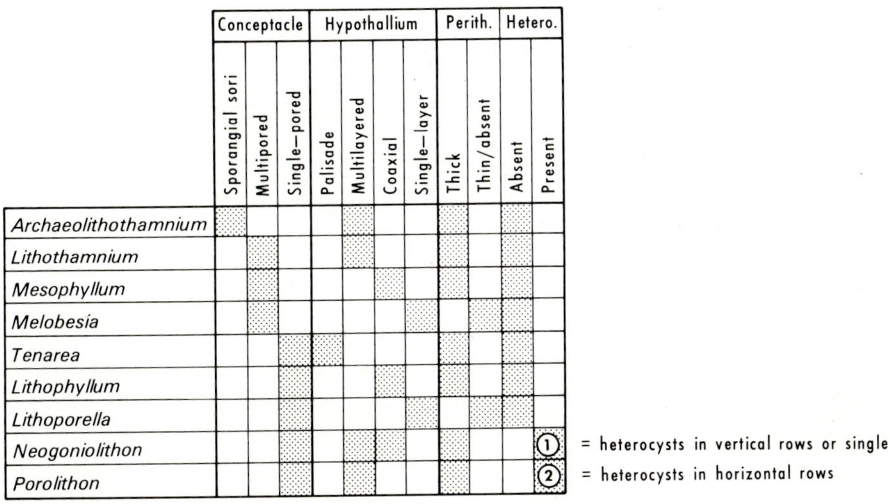

Fig. 48. Morphological characteristics of common genera of crustose coralline algae.

The identification of coralline species is beyond the capabilities of all but specialists. Many morphological variations of living forms are environmentally induced and it is difficult to recognize this influence in fossil populations. As a result, the literature on fossil corallines has been cluttered with many useless species based on superficial characteristics and minor variations in dimensional data.

Crustose coralline algae (Melobesoideae)

This subfamily consists mainly of encrusting genera of variable form and size which inhabit a wide range of marine environments. About 20 living genera belong to the group, and half of these are recognized in the fossil record. In addition, several extinct genera have been described. Some extant genera, not known as fossils, must have existed in earlier times, but have been identified as *Lithothamnium* or other genera because diagnostic features have not been preserved.

Archaeolithothamnium

This genus includes both crusts and branched thalli (Fig. 49) up to several centimeters across. The hypothallium is multilayered, developing parallel to the substrate. Perithallial tissue is generally thick and composed of regular rows of cells. Sporangia are not in conceptacles but occur loose in rows, which is a distinctive characteristic of this alga (Fig. 50).

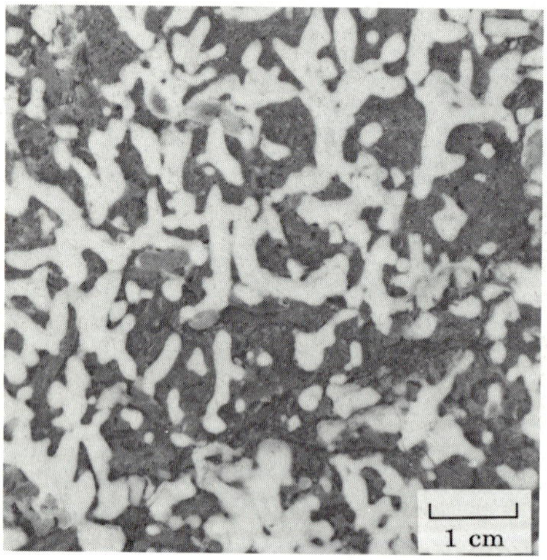

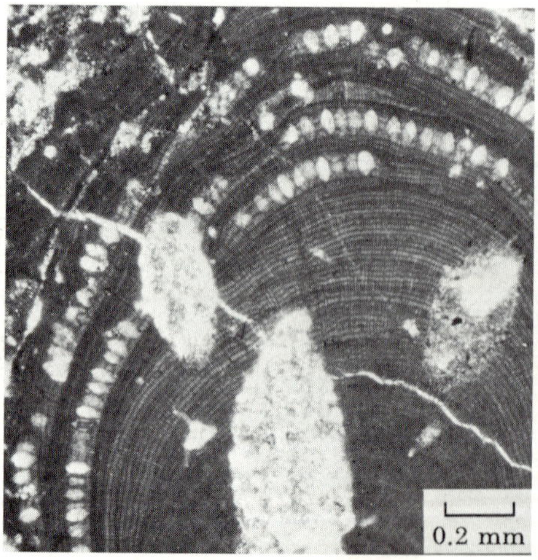

Fig. 49. *Archaeolithothamnium*. Branched growth habit. Paleocene, Libya. Reflected light, polished surface.

Fig. 50. *Archaeolithothamnium*. Perithallial tissue containing rows of individual sporangia. Eocene, Guam. Transmitted light, thin section.

Archaeolithothamnium has been generally considered an ancestral type for all crustose corallines; however, Adey (1970) considered the morphology of *Archaeolithothamnium* inseparable from *Lithothamnium* except for the reproductive structures. The genus may be ancestral to *Lithothamnium* and

Mesophyllum, but it does not appear to be closely related to *Lithophyllum* and *Lithoporella*.

Lithothamnium

Plants occur as crusts and branched forms. The hypothallium is multi-layered, but noncoaxial, developing parallel to the substrate. Perithallium is usually thick and composed of regular layers of cells. Sporangia occur in multipored conceptacles (Fig. 51).

This generic name has been abused in the literature, because it has been used informally to designate many kinds of crustose corallines. On the other hand, some living crustose corallines, for example *Phymatolithon* and *Clathromorphum*, might be identified as *Lithothamnium* in the fossil record, simply because they are basically the same structurally, except for diagnostic non-calcareous cellular features of the meristem and epithallium.

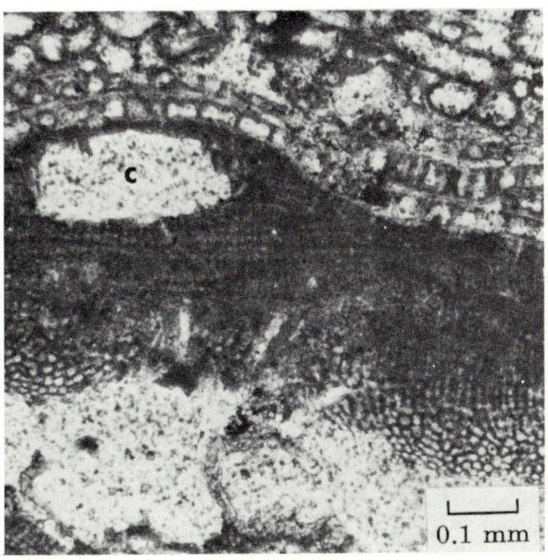

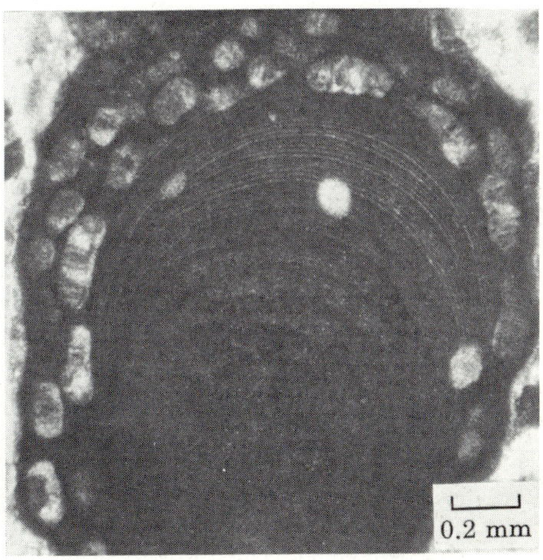

Fig. 51. *Lithothamnium* encrusted by *Lithoporella*. Pleistocene, Caribbean. Transmitted light, thin section.

Fig. 52. *Mesophyllum*. Paleocene, Libya. Transmitted light, thin section.

Mesophyllum

Thalli occur as crusts and some develop branched habits. The multilayered hypothallium is parallel to the substrate and characteristically coaxial (Fig. 52). Perithallium is thick and distinctly layered. Conceptacles are multi-pored and relatively large. This genus is structurally intermediate between *Lithothamnium* and *Lithophyllum*, having conceptacles of the former and cellular tissue similar to the latter. Adey (1972) considered this genus to be morphologically close to *Lithothamnium* and *Archaeolithothamnium*, but not at all near to *Lithophyllum*.

Melobesia

Small, thin crusts composed of a single-layered hypothallium of equidimensional or horizontally elongated cells (Fig. 53) characterize this genus. Perithallium is absent, except around the conceptacles. Conceptacles are multipored and appear as prominent protuberances on the crusts. Thalli are attached to various kinds of substrate and living species are common epiphytes on fleshy algae and marine grasses.

Tenarea

Thalli develop crusts which are attached to a firm substrate or are epiphytic. Hypothallium is composed of a single layer of vertically elongated cells (palisade structure) (Fig. 54). Perithallium is distinctly layered with vertically elongated cells. Conceptacles have a single aperture.

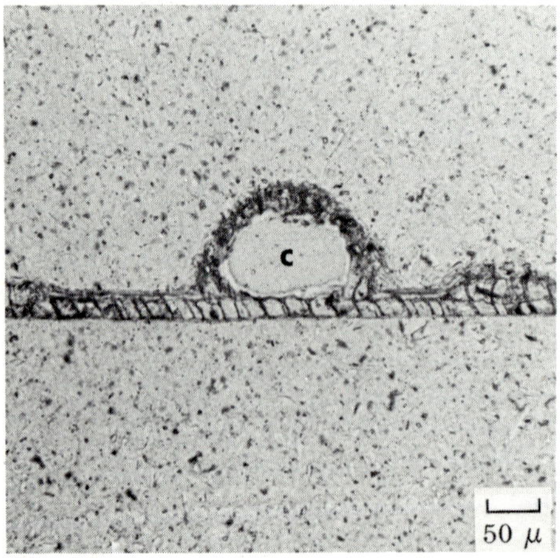

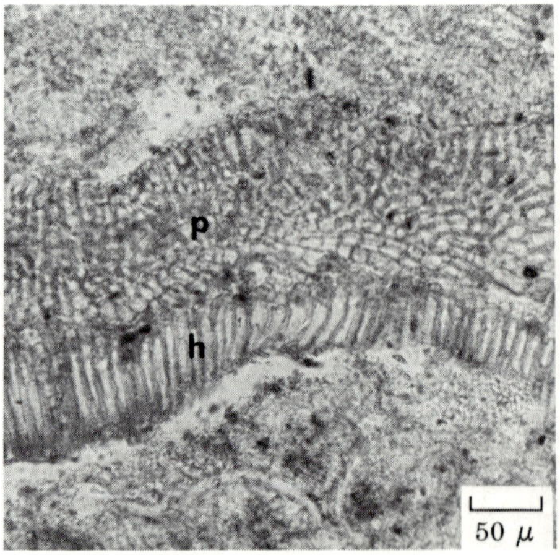

Fig. 53. *Melobesia*. Recent, Florida. Transmitted light, thin section of plastic mount.

Fig. 54. *Tenarea*. Recent, Hawaii. Transmitted light, thin section.

Species of this genus have features of *Lithophyllum* and *Lithothamnium*, and they have also been assigned to *Dermatolithon*. *Tenarea* is a common constituent of Recent algal nodules and coralline pavements known as "trottoirs" in the Mediterranean.

Lithophyllum

This genus occurs most commonly as crusts or branched plants. Characteristically, the hypothallium is coaxial (Fig. 55), but in a few cases it is composed of a single layer of cells. Perithallial tissue is thick, usually composed of regular layers of cells. Branched specimens have a coaxial hypothallium surrounded by a thinner marginal perithallium. Conceptacles have a single aperture (Fig. 46).

Adey (1970) notes that *Lithophyllum*, like *Lithothamnium*, has been used as a "catch-all" genus for various kinds of crustose corallines. Specimens lacking conceptacles are difficult to distinguish from infertile specimens of *Mesophyllum*, because they both have similar cellular tissue and growth habits (Johnson and Adey, 1965).

Secondary pits (Fig. 45) are considered to be a very important systematic character of *Lithophyllum* and *Tenarea*, which sets them apart from other corallines (Adey, 1970). Unfortunately, these very minute structural features cannot be distinguished in thin sections of fossil algae.

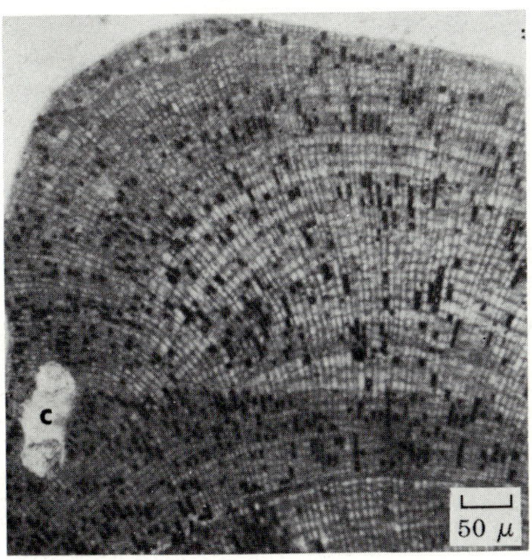

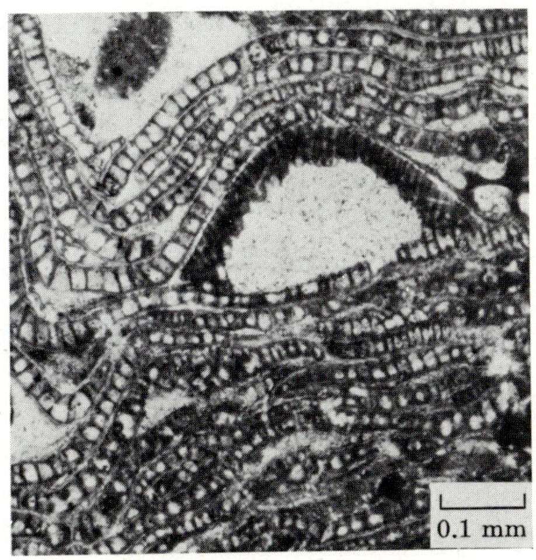

Fig. 55. *Lithophyllum*. Recent, Saipan. Transmitted light, thin section.

Fig. 56. *Lithoporella*. Pleistocene, Caribbean. Transmitted light, thin section.

Lithoporella

This alga develops as a very thin crust, but often thalli overgrow each other repeatedly to form an appreciable composite thickness (Fig. 56). The hypothallium is single-layered and composed of large (up to 40 microns), vertically elongated cells. The perithallium is absent except around conceptacles. Conceptacles have a single aperture.

Identification of *Lithoporella* is rather positive, because of distinctive morphological features. The genus is common and widespread and often encrusts other algae and skeletal constituents.

Neogoniolithon

This genus has both crustose and branched habits. The hypothallium is multilayered and often coaxial; the perithallium is thick and indistinctly layered. The genus is characterized by single or vertical rows of heterocysts within the perithallial tissue (Fig. 57). Conceptacles have a single aperture.

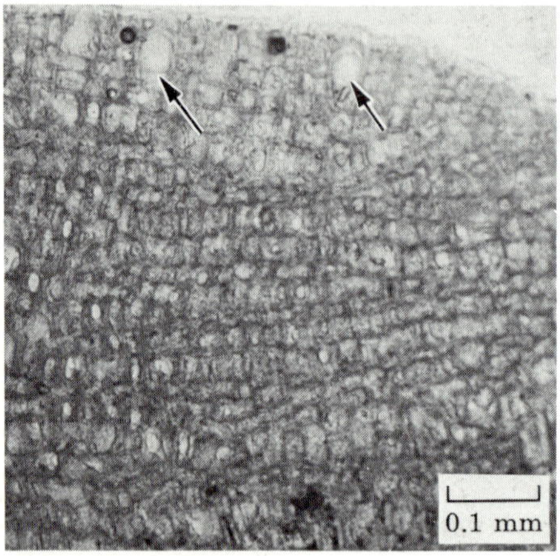

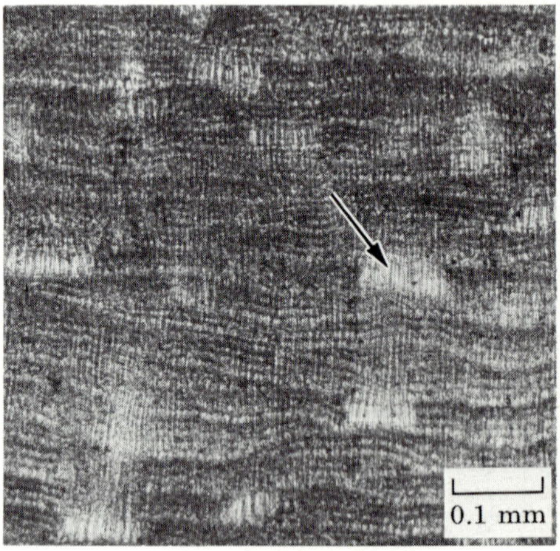

Fig. 57. *Neogoniolithon*. Recent, Pacific. Heterocysts in perithallial tissue. Transmitted light, thin section.

Fig. 58. *Porolithon*. Recent, Pacific. Heterocysts in perithallial tissue. Transmitted light, thin section.

Porolithon

Both massive, compact crustose forms and branched growth habits occur in this genus. The hypothallium is multilayered and composed of cells elongated parallel to the substrate. The perithallium is thick and has distinctly layered cells. Heterocysts occur in horizontal rows within the perithallial tissue (Fig. 58). Conceptacles have a single aperture. The character of heterocysts can be used to distinguish *Porolithon* and *Neogoniolithon*, although other aspects of the structure of the two genera are similar.

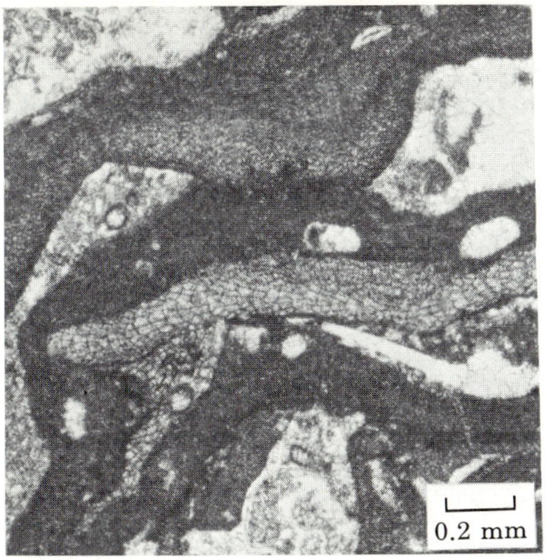

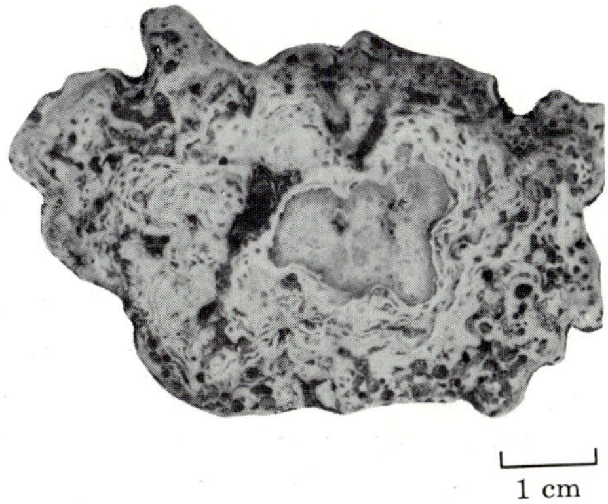

Fig. 59. Crustose corallines. Multiple crusts of *Lithophyllum*. Miocene, Western Australia. Transmitted light, thin section.

Fig. 60. Rhodolith made up of successive layers of corallines. Recent, Bermuda. Reflected light, polished surface. (Courtesy of R. N. Ginsburg).

Articulated coralline algae (Corallinoideae)

About a dozen living genera belong to this subfamily, nearly half of which are known as fossils. In addition, a few extinct genera with limited geologic ranges have been assigned to the Corallinoideae. This morphologically homogeneous group of plants consists of erect, jointed thalli of similar shapes and sizes. Principal structural variations are in details of the cellular tissue and in the position of the conceptacles, which serve as a basis for separation of genera (Johansen, 1969).

Amphiroa

This articulated coralline has segments that vary from cylindrical to flattened forms. The medullary tissue is characterized by one or more rows of long cells alternating with a single row of short cells, and is surrounded by a distinctly layered cortical tissue (Fig. 61). Conceptacles are marginal and lateral in position.

Arthrocardia

Articulated thalli of this genus are characterized by pinnate branching. Segments have straight medullary filaments made up of cells of equal length.

Cortical tissue is relatively thin. Arched layers of cells in the medullary region often appear flattened (Fig. 62). Conceptacles occur in axial positions.

Jania

These plants are characterized by slender, dichotomously branched thalli made up of cylindrical segments. Medullary filaments are surrounded by a thin zone of cortical filaments. Cells of the medullary filaments are typically wedge-shaped and successive rows of cells join along irregular lines (Fig. 63). Conceptacles occur in axial positions.

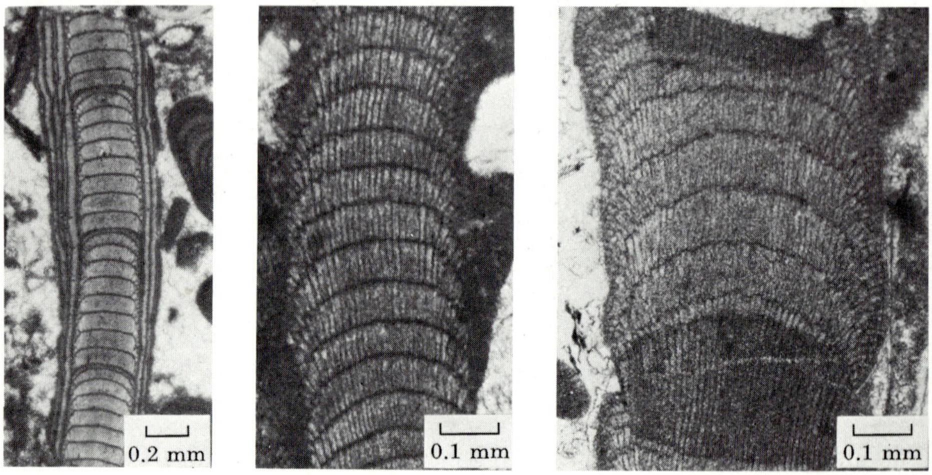

Fig. 61. *Amphiroa*. Medullary tissue bounded by distinctly layered cortical tissue. Pliocene, Pacific. Transmitted light, thin section.

Fig. 62. *Arthrocardia*. Detail of medullary tissue. Eocene, Guam. Transmitted light, thin section.

Fig. 63. *Jania*. Detail of medullary tissue and geniculum. Paleogene, Guatemala. Transmitted light, thin section.

Corallina

The branching habit of this articulated coralline is pinnate, and individual segments vary in shape from cylindrical to clavate. Medullary tissue, composed of regular layers of cells, is dominant in this genus, and cortical tissue is thin and inconspicuous (Fig. 64). Conceptacles occur in axial positions (Fig. 65).

Calliarthron

This genus resembles *Corallina* in its growth habit and general appearance, but differs in that the cellular filaments of the medullary region are flexuous and interlaced (Fig. 66). Conceptacles occur in marginal and lateral positions. This is a common living genus, but is rarely represented in the fossil record.

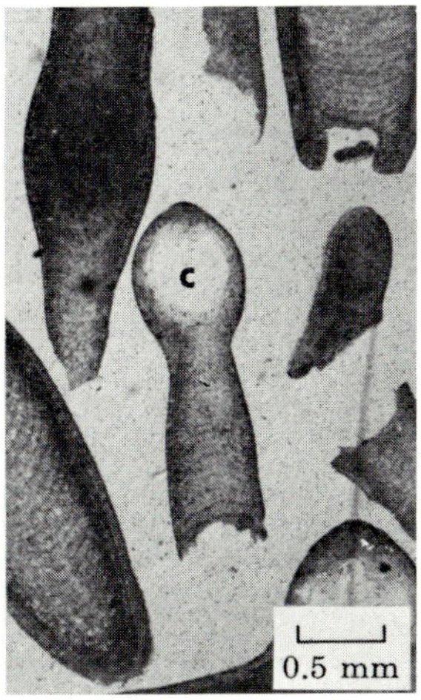

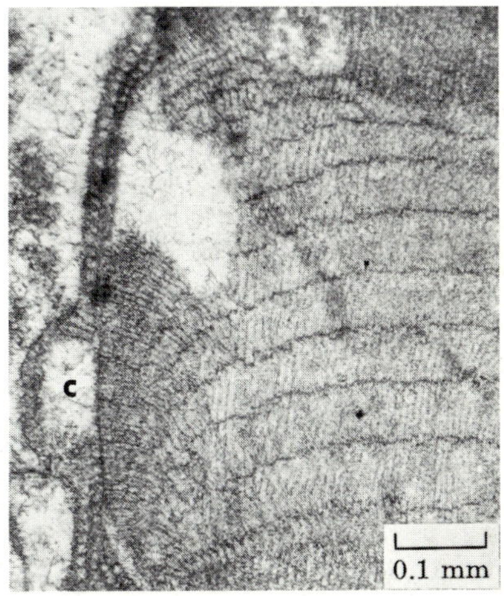

Fig. 64. *Corallina*. Two segments connected by geniculum. Miocene, Pacific. Transmitted light, thin section.

Fig. 65. *Corallina*. Various segments showing cellular tissue and axial conceptacle (c). Recent, California. Transmitted light, thin section of plastic mount.

Fig. 66. *Calliarthron*. Detail of medullary and cortical tissue; conceptacle (c). Miocene, Saipan. Transmitted light, thin section.

Geologic range

The earliest representatives of the Corallinaceae appeared during Jurassic time, and at least seven genera were in existence by the end of the Mesozoic era (Fig. 67). Additional genera have appeared periodically throughout the Cenozoic, but extinctions seem to be rare. The few exclusively fossil genera are poorly known, and so their true distribution in time is uncertain; some of these forms may in fact be synonymous with extant coralline genera.

Adey and Macintyre (1973) outlined a hypothetical evolutionary history of the crustose coralline algae (Fig. 68) utilizing the three subfamilies recognized by Adey and Johansen (1972). All three groups date from the late Mesozoic, but according to these authors the subfamilies are not directly related, having developed from separate ancestors, possibly different genera of Squamariaceae.

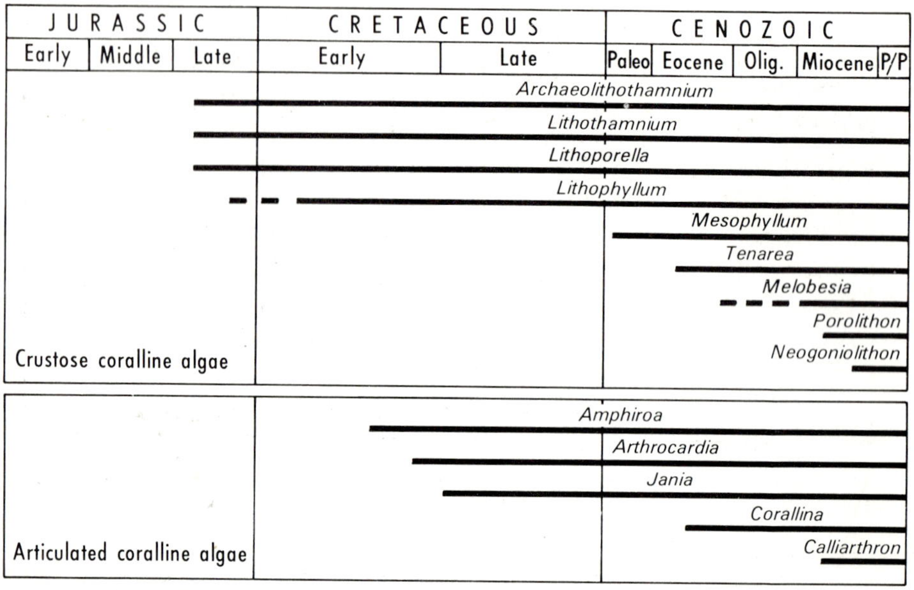

Fig. 67. Geologic ranges of some crustose coralline (Melobesioideae) and articulated coralline (Corallinoideae) genera.

Environmental distribution

Living Corallinaceae are a cosmopolitan, exclusively marine flora. As a group they range from tropical to polar seas and extend from intertidal zones to depths of 250 m. The ecology of living crustose corallines has been summarized by Adey and Macintyre (1973). Principal factors controlling the distribution of the Corallinaceae are listed in Table V.

Temperature is a major factor in the distribution of corallines. Some crustose genera (e.g., *Archaeolithothamnium, Lithoporella, Neogoniolithon,* and *Porolithon*) are restricted to tropical and subtropical waters. *Mesophyllum* and *Lithothamnium* are primarily cold-water genera, however they do occur in lower latitude (warm water) regions in deep environments. *Clathromorphum* is restricted to arctic or subarctic regions. On the specific level, very abrupt cut-off temperatures have been recorded (Adey, 1971), so that temperature is believed to be an important limiting factor.

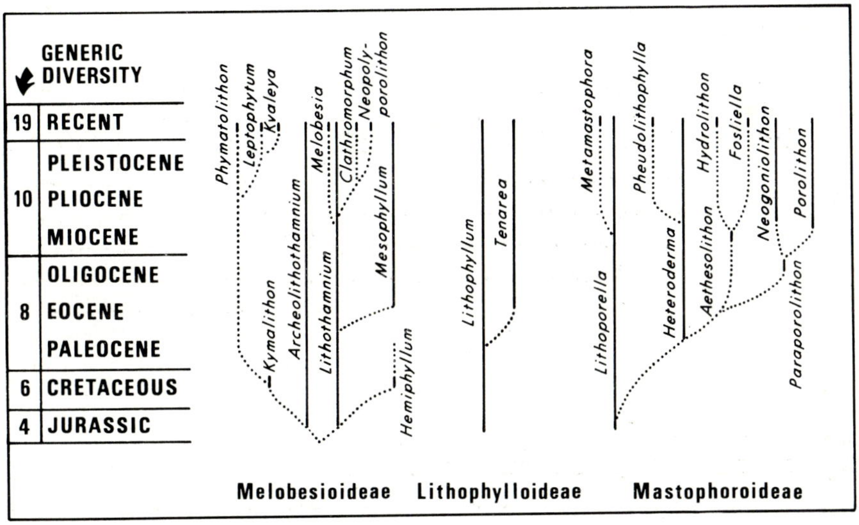

Fig. 68. Suggested evolutionary history of the crustose coralline algae. (After Adey and Macintyre, 1973).

Articulated coralline genera range from tropical to temperate waters, and occur most abundantly at shallow depths of less than 10 m and in the intertidal zone in high energy regimes. *Jania* occurs widely in tropical and subtropical seas. *Corallina* is cosmopolitan in tropical, subtropical, and temperate waters, and a few species approach subarctic regions. Most other articulated genera are prevalent in subtropical and temperate environments.

According to Adey and Adey (1973), light is the primary factor controlling the depth distribution of crustose corallines. Because some members of the Corallinaceae are able to survive in very low-intensity light conditions, they may extend to great depths or occur in concealed situations in shallow waters. Considered as a group, the corallines have an exceedingly broad depth distribution, but most species have limited depth ranges. At the generic level, depth zonation is less sharply delineated, yet most genera have characteristic ranges of occurrence (Fig. 69).

Both crustose and articulated coralline groups occur principally in waters of normal marine salinities. A few crustose species inhabit brackish-water environments, but these are rather minor occurrences.

The character of the substrate and wave energy are factors in controlling distribution, and the corallines display a wide range of requirements.

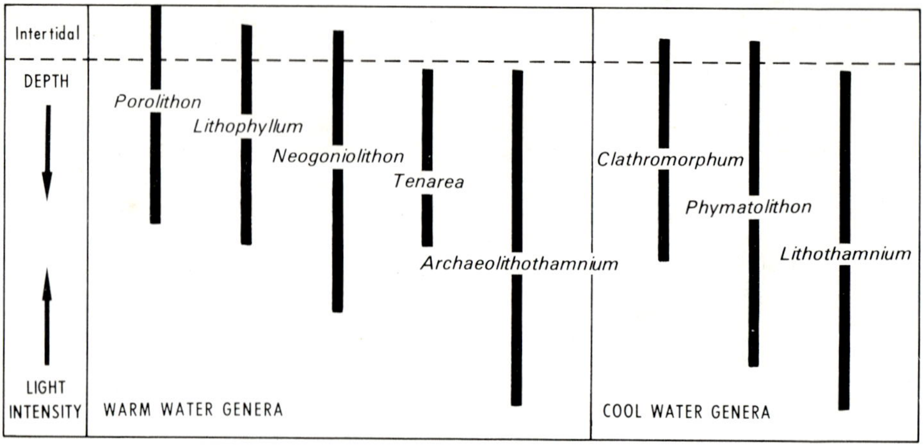

Fig. 69. Generalized depth distribution and light requirements of some living crustose coralline genera. (After Adey and Macintyre, 1973).

A fixed or firm substrate is essential to many taxa, including all articulated genera, whereas others tolerate unstable, soft bottom habitats.

The Corallinaceae are important Cenozoic reef builders in tropical and subtropical realms, both as framebuilding organisms and sediment producers. The role of corallines in reef construction is quite apparent in most existing reefs. Littler (1973) determined that crustose coralline algae cover nearly 40 percent of the reef surface of some Hawaiian fringing reefs. Scleractinian corals are volumetrically significant in most living reefs, but crustose corallines are essential binding and cementing agents. Predominant coralline components of tropical reefs today are *Neogoniolithon, Porolithon,* and *Lithoporella*. *Mesophyllum* and *Archaeolithothamnium* are less important, and *Lithothamnium* is absent or relatively unimportant (Adey and Macintyre, 1973).

Crustose corallines develop laminated nodules or rhodoliths (Bosellini and Ginsburg, 1971) in some unstable substrate environments (Fig. 60). A delicate balance between water motion and light conditions, which permits essentially continuous growth on all surfaces of the nodule, is required for rhodolith development. According to Adey and Macintyre (1973), branched species of *Lithothamnium* are the dominant rhodolith forms, although *Archaeolithothamnium, Lithophyllum* and *Neogoniolithon* are also contributors to the construction of these kinds of algal structures.

TABLE V

FACTORS CONTROLLING DISTRIBUTION OF LIVING CORALLINACEAE

TEMPERATURE	Tropical to arctic/antarctic regions; limited temperature ranges at generic and specific levels.
DEPTH	Intertidal to about 250 m; restricted depth zonation at generic and specific levels.
SALINITY	Mostly normal marine; few species tolerate lower salinities in coastal habitats.
SUBSTRATE	Variable; most taxa require attachment to firm substrate, others prefer soft or unstable bottoms.
ENERGY	Wide variation; some taxa require intense wave agitation or water motion, others live in very low energy regimes.

OTHER CALCAREOUS RED ALGAE -- REAL AND PROBLEMATICAL

Numerous calcareous fossils, mostly Paleozoic taxa, are believed to belong to the red algae; however, their morphologies are not similar enough to living algae to positively assign them to extant families. Some forms have definite characteristics of modern red algae, but uncertain affinities within the phylum, while others have a problematical relationship to the Rhodophyta. Despite these taxonomic uncertainties, fossils presumed to be calcareous red algae are diverse and widespread constituents in Paleozoic carbonate rocks, and therefore are useful in paleoenvironmental reconstructions and biostratigraphy.

Ancestral corallines

Several Late Paleozoic genera closely resemble members of the Corallinaceae, but these fossils have not been definitely linked to this family because of structural differences and the significant time hiatus between their last appearance in the Permian and the first occurrence of the true corallines in the Jurassic. Nevertheless, they probably represent an ancestral stock to the modern Corallinaceae.

At the present time, two ancestral coralline genera, *Archaeolithophyllum* and *Cuneiphycus*, are known to be abundant and widespread. Two others of coralline character, *Katavella* and *Lysvaella*, are based on limited numbers of specimens from a few localities. According to Chuvashov (1965), the Devonian genus *Katavella* has an unsegmented cylindrical thallus consisting of differentiated tissue of prismatic cells and external reproductive organs. *Lysvaella* (Chuvashov, 1971) from the Permian of the U.S.S.R. has tissue differentiated into hypothallium and perithallium, in addition to reproductive bodies embedded within the perithallial tissue, characteristics which resemble *Archaeolithophyllum*. Enough data may be assembled eventually to establish the taxonomic affinities of these Paleozoic algae.

These kinds of fossils have been misinterpreted, and caution is necessary in attempts to establish biological relationships. For example, the Devonian genus *Keega* from Western Australia was originally described as a red alga of uncertain affinities, but probably related to the crustose Corallinaceae (Wray, 1967). The fossil was believed to have differentiated tissue and reproductive organs similar to *Lithophyllum*, although a subsequent study of additional specimens from Alberta (Wray and Playford, 1970) cast some doubt on the algal affinity of *Keega*. In 1974 Riding reinterpreted the fossil as a stromatoporoid, specifically the basal layer of a laminar form, as opposed to the more common dendroid form, of the genus *Stachyodes*. Riding demonstrated that the internal structure, or "cellular tissue", of *Keega* is a combination of primary megastructure and diagenetically altered microstructure, and the "conceptacles" are fortuitous sections of tubes which form branched networks within the stromatoporoid.

Archaeolithophyllum

This alga consists of a thoroughly calcified, irregularly shaped, crustose thallus several centimeters in length and 0.2-0.8 mm thick. Plants occur as solitary or multiple crusts, and as foliate masses (Figs. 70, 71). Some were attached during growth, others apparently developed free on the substrate (Johnson, 1956; Wray, 1964).

Cellular tissue is differentiated into a thick medullary hypothallium and a thin perithallium (Fig. 72). The hypothallium is composed of arcuate rows of large polygonal cells up to 150 microns in length; considerably larger than those in modern corallines. The perithallium is made up of smaller, rectangular cells arranged in rows parallel to the surface of the thallus. Subconical conceptacles with a single aperture are distributed irregularly over the upper surface of the thallus (Fig. 73).

Archaeolithophyllum ranges from the Early Carboniferous (Late Mississippian) to Late Permian, and is geographically widespread in shelf carbonate

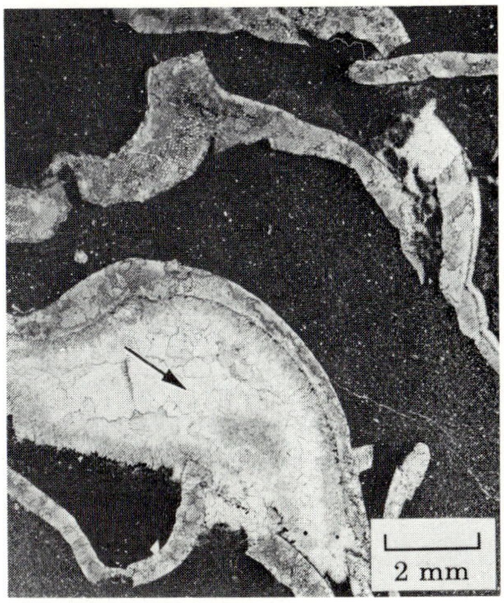

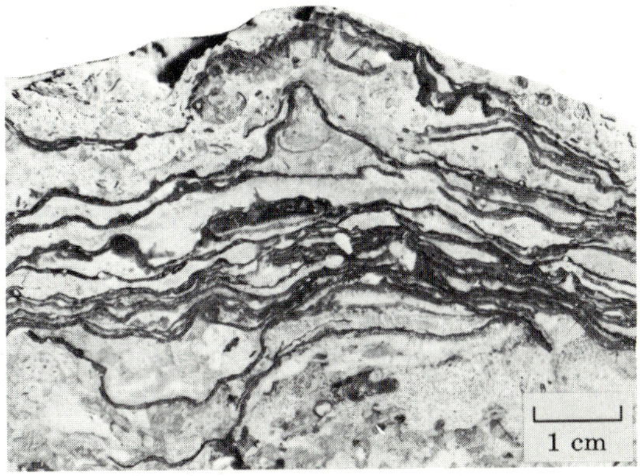

Fig. 70. *Archaeolithophyllum*. Vertically oriented section of specimens. Note sparry calcite filling of original void beneath crust. Upper Carboniferous (Pennsylvanian), Kansas. Transmitted light, thin section.

Fig. 71. *Archaeolithophyllum*. Vertically oriented surface of limestone showing multiple crusts. Upper Carboniferous (Pennsylvanian), Kansas. Reflected light, polished surface.

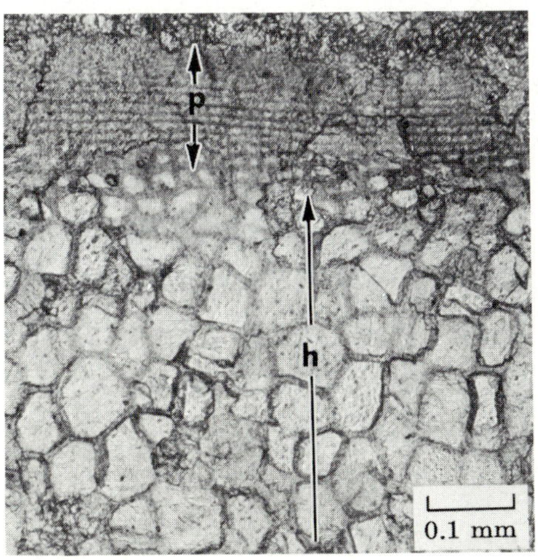

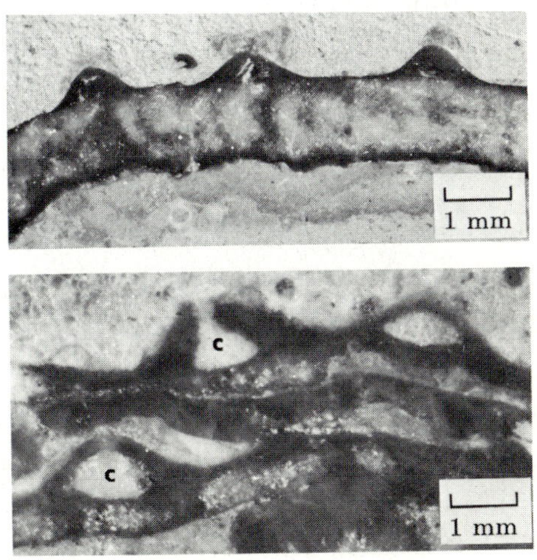

Fig. 72. *Archaeolithophyllum*. Detail of cellular tissue; hypothallium (h) and perithallium (p). Upper Carboniferous (Pennsylvanian), Kansas. Transmitted light, thin section.

Fig. 73. *Archaeolithophyllum*. Transverse section of three conceptacles (above). Multiple crusts with conceptacles (below). Upper Carboniferous (Pennsylvanian), Kansas. Reflected light, polished surfaces.

deposits of North America, Europe, and the U.S.S.R. This alga may be the dominant fossil of carbonate facies, occurring to the exclusion of other skeletal constituents.

Cuneiphycus

This genus has a rigid, branched thallus extending for a few centimeters, but it is not known whether it was attached or developed free. Originally, this alga was believed to have a cylindrical, segmented thallus, because it was found only as fragments which resembled articulated coralline algae (Johnson, 1960). However, more complete specimens show complex three-dimensional, chainlike networks (Fig. 74). Internally, the genus is characterized by a simple, undifferentiated tissue of large rectangular or wedge-shaped cells up to 100 microns long (Fig. 75). Reproductive organs are unknown.

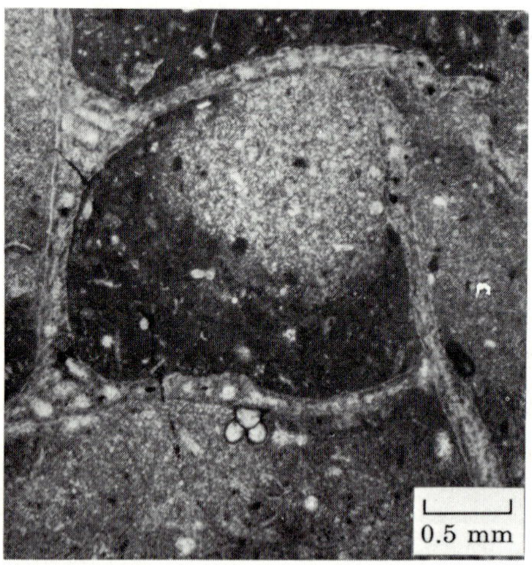

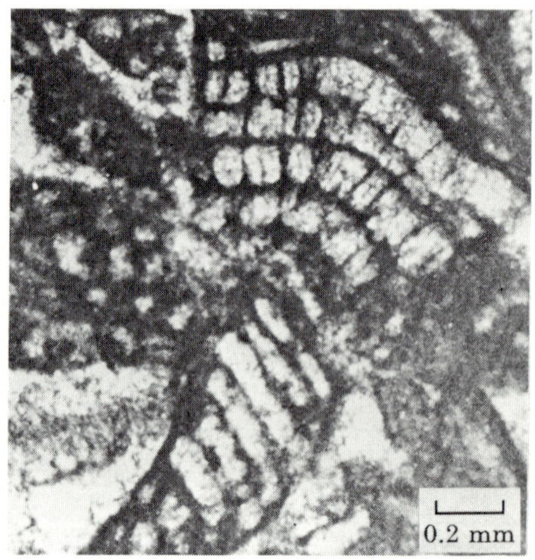

Fig. 74. *Cuneiphycus*. Branched crustose thallus. Upper Carboniferous (Pennsylvanian), Oklahoma. Transmitted light, thin section.

Fig. 75. *Cuneiphycus*. Detail of cellular tissue. Upper Carboniferous (Pennsylvanian), Texas. Transmitted light, thin section. (From Johnson, 1960).

Cuneiphycus has been described from the Late Carboniferous (Pennsylvanian) in North America and Europe. Locally it occurs commonly in carbonate bank accumulations.

Problematical red algae

The biological affinities of these taxa have been the subject of much unresolved controversy. Many of the names have long been associated with the red algae, yet the morphological evidence for this relationship is suspect. Although some problematical red algae have been described from the Mesozoic, for example *Polygonella* (Elliott, 1957), the majority are Paleozoic fossils, some of which are abundant and widespread.

The Carboniferous genus *Litostroma* (Mamet, 1959) and the Devonian genus *Stenophycus* (Wray, 1967) have been classified as red algae of uncertain affinities. Both genera have a peculiar cellular tissue and each is known only from a few localities; their relationship to the red algae remains unclear.

Komia et alia

Taxa considered here include *Ungdarella, Komia, Foliophycus, Stacheia, Stacheoides,* and *Aoujgalia*. These genera are usually treated as a related group, although the biological affinities of these fossils continue to be questioned. Also, a few genera may be synonymous with others within the group. Some have a branched habit (Fig. 76), while others are irregular, encrusting forms. All were attached and are characterized by a fibrous wall structure (Fig. 77). They lack a distinctly cellular tissue, and have no preserved reproductive organs. These fossils have at one time or another been assigned to the red algae in the family Ungdarellaceae (Maslov, 1956).

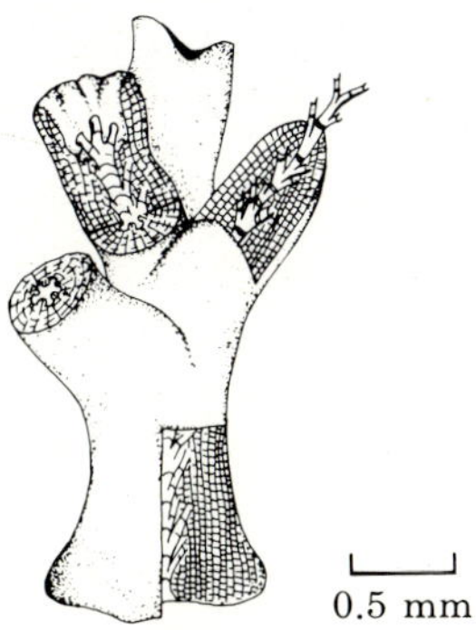

Fig. 76. *Komia*. Reconstruction of growth habit and internal structure. (Fisher, 1975).

These fossils seem to be divisible into two groups. One group includes *Stacheia*, *Stacheoides*, and *Aoujgalia*; and the other, *Ungdarella*, *Komia*, and *Foliophycus*. The *Stacheia* group is restricted to the Early Carboniferous and was succeeded by the *Ungdarella-Komia* group in Middle Carboniferous time. Stacheiinids have been classified in the foraminifera, although the wall structure is not typical of Paleozoic foraminifers; thus they are often placed in the red algae (e.g., Petryk and Mamet, 1972). On the other hand, Wilson et al., (1963) and Wilson (1969) concluded that *Ungdarella* and *Komia* are stromatoporoids. More recently, Fisher (1975) provisionally assigned *Komia* to the hydrozoans, and all of these genera, along with others, are considered to be calcareous sponges by Termier et al., (1975).

Despite the large amount of disagreement concerning the classification of these fossils, some forms are common constituents in Late Paleozoic carbonate sediments and occur in biohermal accumulations. Many have restricted geologic ranges, and therefore are useful in stratigraphic correlations and age determinations.

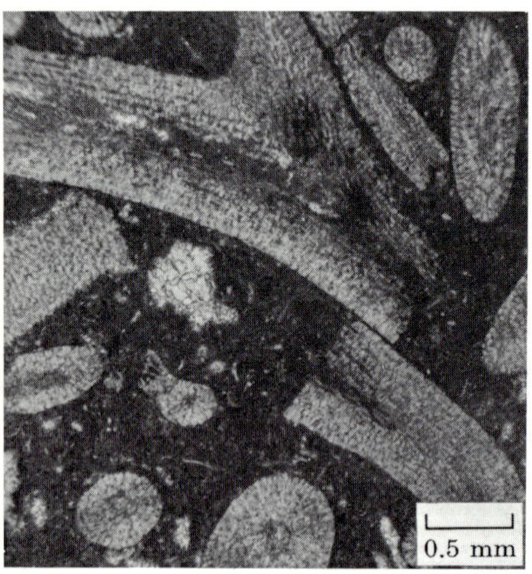

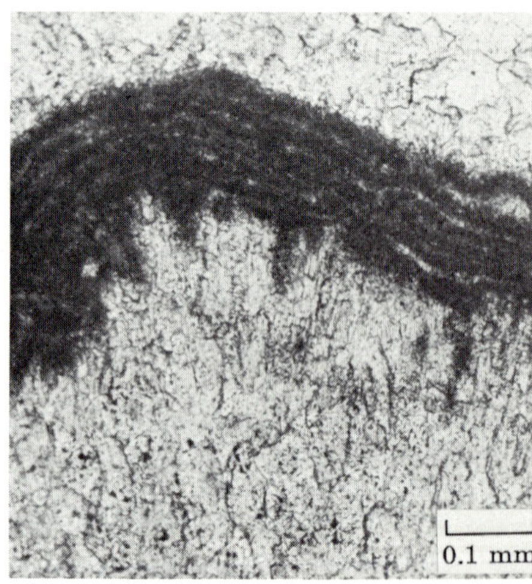

Fig. 77. *Komia*. Variously oriented sections. Upper Carboniferous (Pennsylvanian), Colorado. Transmitted light, thin section.

Fig. 78. *Archaeolithoporella*. Encrusting laminae. Permian, Texas. Transmitted light, thin section.

Archaeolithoporella

This Permian genus was originally described by Endo (1959) as having a morphology similar to *Lithoporella*. However, based on a reexamination of the holotype and study of numerous Late Permian specimens, Babcock (1974) demonstrated that *Archaeolithoporella* has a very simple, encrusting, laminar structure without cells (Fig. 78). Consequently, there seems to be no evidence to justify classification of this fossil among the coralline red algae; however, it is considered to be of organic origin and probably algal. This is an important reinterpretation, because the generic name has become entrenched in the literature as an early fossil crustose coralline alga.

Geologic range and environmental distribution

Several taxa have relatively limited geologic ranges and are useful index fossils (Fig. 79). Many Carboniferous genera are geographically widespread, having been reported from numerous regions.

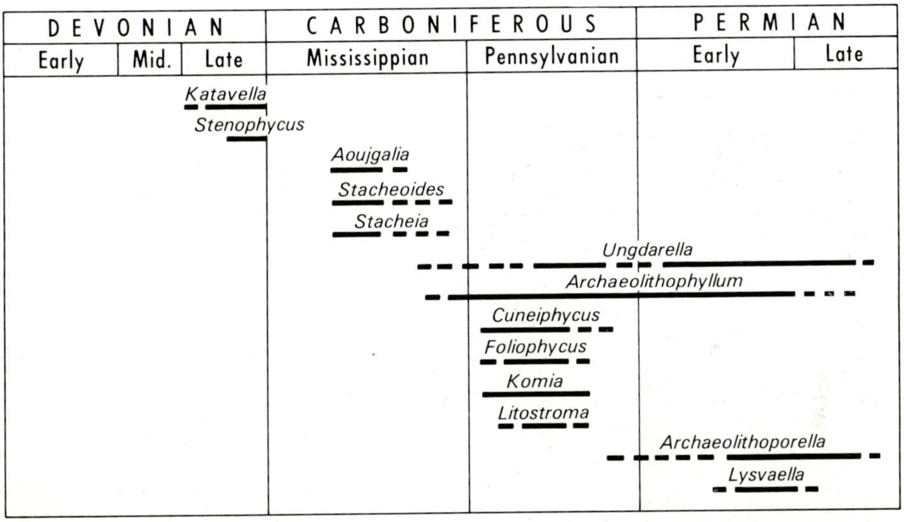

Fig. 79. Geologic ranges of late Paleozoic problematical calcareous red algae.

Komia and related forms are important contributors of skeletal grains to various kinds of carbonate facies. *Archaeolithophyllum* is the dominant constituent in some Upper Carboniferous and Lower Permian carbonate banks and reefs. These genera and other problematical red algae inhabited shallow-shelf environments.

Chapter 5

CALCAREOUS GREEN ALGAE (CHLOROPHYTA)

The Chlorophyta comprise a large group of algae that includes forms with diverse lines of development and a wide range of morphologies inhabiting mainly terrestrial and freshwater environments. Marine forms make up less than 15 percent of the total flora. Due to the predominance of green pigments, most Chlorophyta have a distinctive grass-green color, except where it is masked by calcification.

Calcareous green algae belong to the following three groups. The first two listed are families and inhabit exclusively marine environments; the charophytes are nonmarine and have been variously considered a separate division (phylum), as well as different ranks within the Chlorophyta.

- Codiaceae
- Dasycladaceae
- Charophyceae

Marine green algae are primarily macroscopic plants with well defined cell walls. Portions of the thallus are impregnated or coated with calcium carbonate in some members of the Codiaceae and in all of the Dasycladaceae. Specialized calcification of the reproductive organs take place in the Charophyceae. Devonian noncalcareous green algae belonging to the order Volvocales were described by Kazmierczak (1975) and their preservation attributed to early diagenetic precipitation of calcium carbonate.

CODIACEAE

This family, one of seven belonging to the order Siphonales, is the only one of the order in which species precipitate skeletal carbonate. Although the Siphonales vary greatly in shape and size, thalli of all forms are composed of filamentous and tubelike (siphonaceous) structures.

Characteristics

Seventeen living genera belong to the Codiaceae (Dawson, 1966), but only five of them are calcified. Calcareous forms are erect plants attached by rhizoids and exhibit variations in external morphology ranging from dense, compact thalli of *Halimeda* (Fig. 2) to delicately branched and loosely calcified genera, such as *Penicillus* and *Udotea*. These variations are reflected in the skeletal content ($CaCO_3$) expressed in percentage of the dry weight of the entire plant: *Halimeda*, 97 percent; *Penicillus* and *Rhipocephalus*, 60 percent; and *Udotea*, 37 percent (Böhm, 1973).

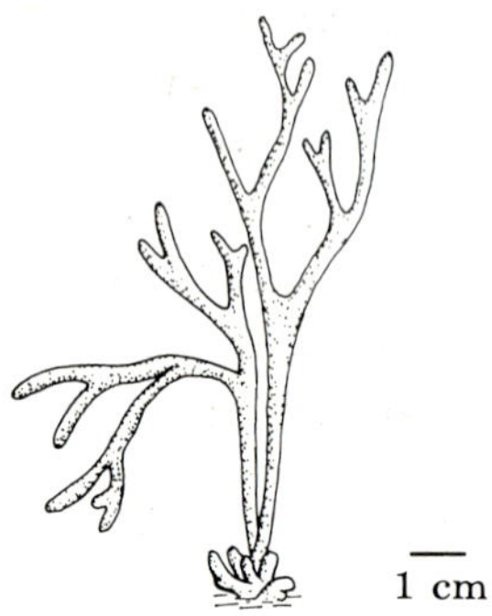

Fig. 80. *Codium*. A living noncalcareous codiacean with an erect growth habit.

The internal structure of all genera is characteristically composed of a central region (medulla) of branched and interwoven filaments and a peripheral layer (cortex) of utricles (Fig. 5). Aragonite is precipitated within interfilament regions of the thallus.

Among living calcareous codiacean genera, only *Halimeda* is preserved as recognizable skeletal grains in the sedimentary record; the others disaggregate into unidentifiable, fine-grained aragonite crystals. Both types of particles are important sources of carbonate sediments in low-latitude, shallow-shelf regions.

Classification

Fossil calcareous Codiaceae display variations in external and internal morphologies comparable to living calcified genera; thus similar taxonomic critera may be used. Only fossil codiaceans that grew as erect plants are considered; nodular and encrusting forms, such as *Ortonella* and *Cayeuxia*, previously classified as Codiaceae, are assigned to the calcareous blue-green algae. Some 15-20 fossil genera with erect growth habits have been described, but the affinity of a few is problematical. Generic criteria for classifying representatives of this family include:

- Growth form: nonsegmented, segmented, and bladed.
- Internal variations in the size, shape, and arrangement of medullary and cortical filaments.
- Reproductive organs.

The species concept in the Codiaceae was reviewed by Konishi (1961), who considered both biological and paleontological characteristics. Despite the common nature of living calcareous codiaceans, very little detailed taxonomic work has been done. An exception is the monographic study and revision of *Halimeda* by Hillis (1959), in which she recognized 21 species. Some species may be distinguished on the basis of shape and size of segments, without resorting to differences in internal morphology. By this method Wiman and McKendree (1975) determined the distribution of disarticulated segments of seven species and correlated them with the living *Halimeda* flora.

Representative genera

Palaeoporella

Palaeoporella seems to be the oldest codiacean green alga, presumably ranging from the latest Cambrian into the Devonian (Johnson, 1966). First described by Stolley (1893), this genus consists of an unsegmented, cylindrical thallus which bifurcates in the upper portion. Specimens are several centimeters long and 1-2 mm in diameter. The growth form of *Palaeoporella* appears to have been similar to the Recent uncalcified genus *Codium* (Fig. 80).

Internally, the structure is typically codiacean, being composed of longitudinally arranged medullary filaments that turn towards the exterior and branch within the cortical region (Fig. 81). Silicified specimens in Ordovician erratic boulders from Poland have provided the most complete understanding of the external and internal morphology of *Palaeoporella*, and confirmed the assignment of this genus to the Codiaceae (Kozolowski and Kazmierczak, 1968).

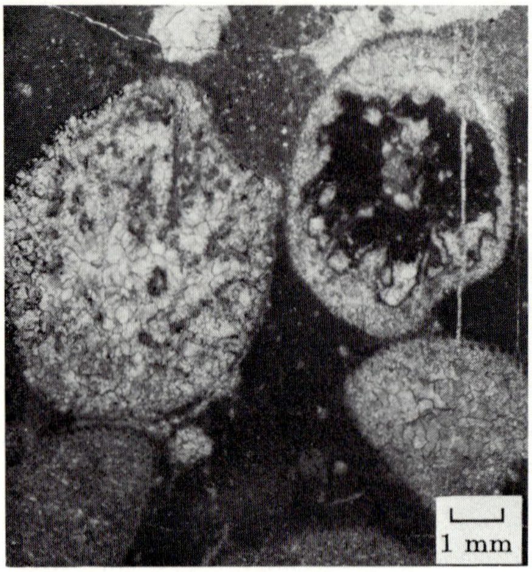

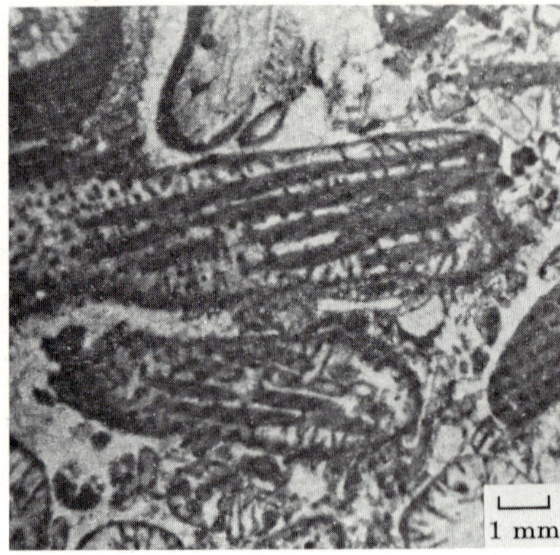

Fig. 81. *Palaeoporella*. Ordovician, Norway. Transmitted light, thin section.

Fig. 82. *Dimorphosiphon*. Longitudinal and transverse sections of segments. Ordovician, Norway. Transmitted light, thin section. (From Høeg, 1927).

Dimorphosiphon

The thallus of this Ordovician genus (Høeg, 1927), in contrast to *Palaeoporella*, is segmented like *Halimeda*. *Dimorphosiphon* has cylindrical, calcified segments about 1 cm long and average 2.5 mm in diameter. The internal differentiation of filaments of this genus is considered to be rather simple, consisting of large medullary tubes and simple cortical filaments (Fig. 82).

Litanaia

This genus has a similar morphology to *Abacella* and *Lancicula* and all three are restricted to the Devonian (Maslov, 1956; Johnson, 1964; and Wray, 1967). The thallus of *Litanaia* consists of cylindrical, calcified segments about 1-2 mm in diameter and up to 4 mm long. The medullary region is

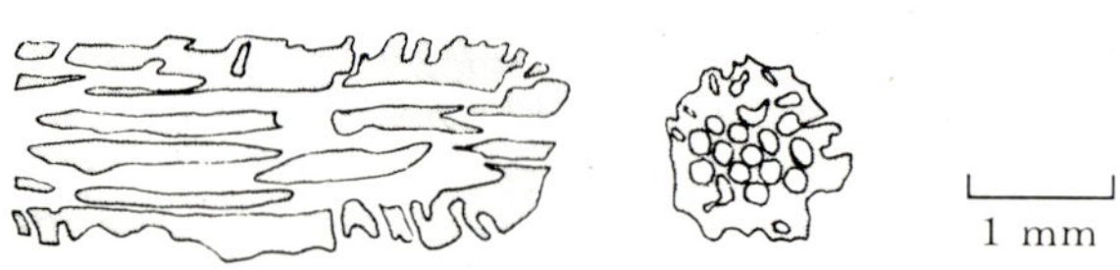

Fig. 83. *Litanaia*. Longitudinal (left) and transverse (right) sections. Devonian, Western Australia.

composed of a bundle of coarse tubular filaments arranged parallel to the long axis of the thallus. The cortex is perforated by finer tubes arranged at high angles or nearly perpendicular to the medulla (Fig. 83).

Representatives of these genera appear to have been geographically widespread during the Devonian, occurring in Siberia, eastern and Western Australia, and various localities in North America.

Eugonophyllum and Ivanovia

These two genera, along with *Anchicodium* and *Calcifolium*, are similar morphologically, and some of them may be synonymous. The thallus of all four consists of a thoroughly calcified blade (or blades) of variable shape, comparable to the form of the Recent codiacean genus *Udotea*. Late Carboniferous silicified specimens suggest that the shape of *Eugonophyllum* may have resembled the growth form of some Recent noncalcareous brown algae (Fig. 85). Fragments of these skeletal plates measure several centimeters long and 0.5-1.0 mm thick. Entire plants may have reached 10 cm or more in height.

Internally, blades may be simple monostromatic (one-cell thick) arrangements, as in *Calcifolium* (Maslov, 1956), or differentiated into medullary and cortical regions, as in *Anchicodium, Ivanovia,* and *Eugonophyllum* (Konishi and Wray, 1961) (Fig. 84). The Middle Carboniferous genus *Calcifolium* has

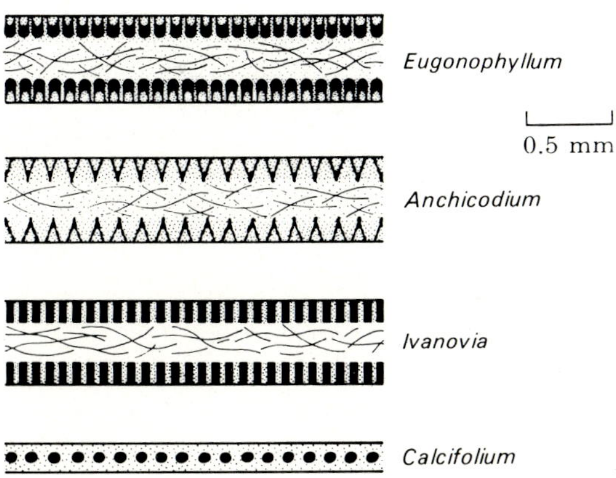

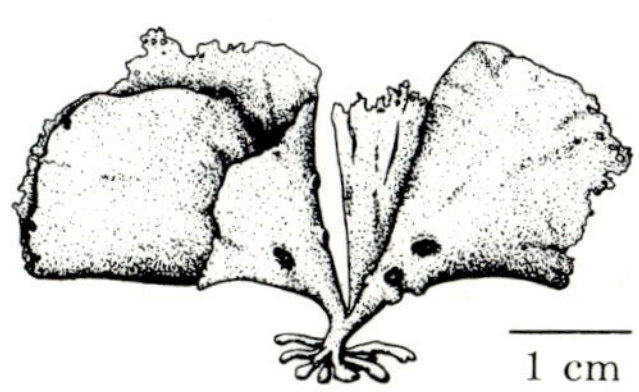

Fig. 84. Diagrammatic transverse sections of bladed Late Paleozoic Codiaceae. Utricles shown in solid black; degree of calcification indicated by intensity of stippled pattern.

Fig. 85. Suggested growth habit of *Eugonophyllum*.

dichotomously branched filaments arranged in the same plane. The cortical structure of *Eugonophyllum* is differentiated into two layers, whereas *Anchicodium* and *Ivanovia* have a single layer. These subtle differences may be artifacts due to diagenesis; however, a basic characteristic of *Eugonophyllum*, namely reproductive organs borne within subcortical protuberances, has not been reported in the other genera (Fig. 86).

Diagenesis has often altered the internal structure to mosaics of sparry calcite. In these cases it is virtually impossible to distinguish one genus from another, and from the ancestral coralline alga *Archaeolithophyllum* which is often associated with these codiaceans. Pray and Wray (1963) suggested the term "phylloid" be applied to all of these Late Paleozoic calcareous algae with similar leaflike forms when they lack sufficient internal features for generic identification. *Eugonophyllum* and related codiaceans are common fossils. They occur worldwide, and may be the dominant skeletal constituent in mounds and other carbonate facies of Late Carboniferous and Early Permian age.

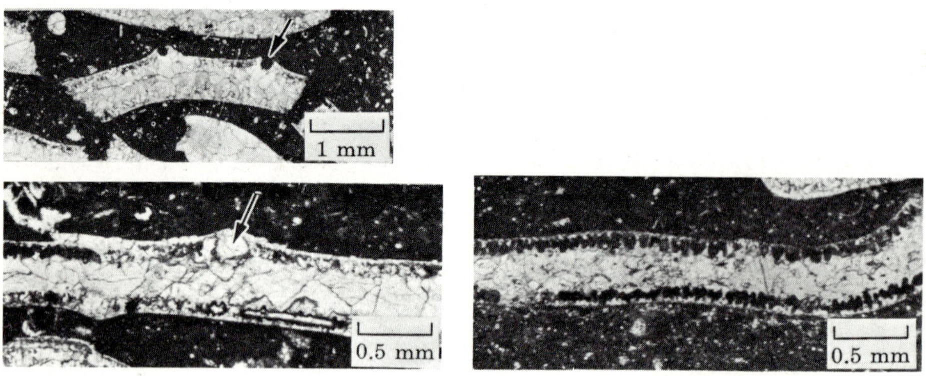

Fig. 86. *Eugonophyllum*. Transverse sections; reproductory organs (upper and lower left) and cortical structure (right). Upper Carboniferous (Pennsylvanian), New Mexico. Transmitted light, thin section.

Hikorocodium

This cylindrical, segmented genus is differentiated internally into medullary and cortical regions (Fig. 87); however, the structure of the central stem is poorly organized and rarely preserved. The cortex is thoroughly calcified but molds of the cortical filaments seem to lack detail. Despite its vague internal morphology, *Hikorocodium* is a distinctive fossil, presumably belonging to the Codiaceae. This genus, first described by Endo (1951) from the Permian of Japan, has been recorded subsequently in Europe and North Africa, and ranges from the Late Carboniferous through the Permian.

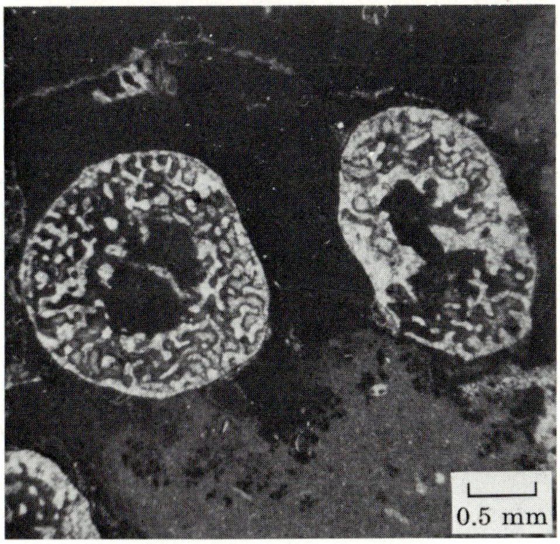

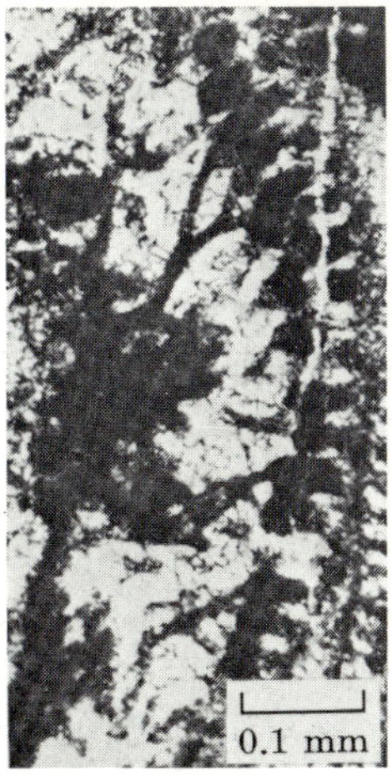

Fig. 87. *Hikorocodium*. Longitudinal and transverse section. Permian, Tunisia. Transmitted light, thin section.

Fig. 88. *Succodium*. Longitudinal section; detail of cortex. Permian, Japan. Transmitted light, thin section. (From Konishi, 1954).

Succodium

Succodium is a cylindrical, segmented Permian codiacean (Konishi, 1954). It has an advanced internal morphology consisting of medulla, a distinctive subcortex of inflated utricles, and an outer cortex with much finer filaments (Fig. 88).

Ovulites

Ovulites was one of the first fossil calcareous algae described (Lamarck, 1816), but it was not considered to be analogous to Recent Codiaceae until much later (Munier-Chalmas, 1879). Originally thought to be restricted to the Eocene, the earliest occurrence of the genus has been extended to the Late Cretaceous.

Ovulites grew as an erect, articulated plant several centimeters high, consisting of segments varying in shape from simple cylinders to globular forms up to 2 mm long and generally less than 0.5 mm in diameter (Fig. 89). Specimens are hollow, suggesting that the medullary region was uncalcified, but a finely perforate network is preserved in the cortex.

Halimeda

This heavily calcified, segmented codiacean is a well known marine calcareous alga. *Halimeda* has been a prodigious sediment producer throughout its history which extends from the Mesozoic. Individual plants grow erect, although some may be draped like vines on steep slopes in deep waters. Calcified segments, varying in shape from flattened to subcylindrical forms, are separated by weakly calcified nodes, and upon death, whole plants generally disarticulate into individual segments. Internally, they consist of longitudinal filaments in the medullary region that in turn develop lateral filaments in the cortex where the terminate in a surface layer of cortical utricles (Fig. 91).

Calcification begins at the exterior of a segment and proceeds inward; also, older segments are more thoroughly calcified than younger ones. The aragonitic mineralogy of the thallus invariably alters to sparry calcite in fossils, but retains the internal morphology of the segments. Reproductive organs (sporangia) of *Halimeda* occur as noncalcareous outgrowths of the segments and are not preserved.

The first occurrence of *Halimeda* is not certain because of taxonomic problems in distinguishing the morphologically similar Jurassic and Early Cretaceous genera, *Boueina* and *Arabicodium*. These two Mesozoic forms typically have cylindrical segments, but the internal structures of all three show overlapping characteristics. Elliott (1965) and Johnson (1968) concluded that these codiacean genera should be referred to the single genus *Halimeda*, which has priority. If this usage is followed, then Middle Jurassic time is the earliest appearance of this modern calcareous green alga.

Problematical codiacean algae

Microcodium

Microcodium, a problematical microorganism, has been associated with the Codiaceae since it was first described by Glück (1912). Johnson (1961) was quite convinced of its affinity to the green algae, but Pia (1927) and Konishi (1961), among others, doubted that it was a codiacean alga, and some authors have suggested an inorganic (diagenetic) origin.

The genus can best be described as groups of elongate, radiating crystals (approximately 1 mm long) which occur in subcylindrical or sheet-like clusters (Fig. 93). First described from Miocene rocks, it is now believed to range from the Late Jurassic to Recent.

Estaban (1974), in a review of *Microcodium* and examination of Eocene specimens of Spain, found this form directly involved in caliche and

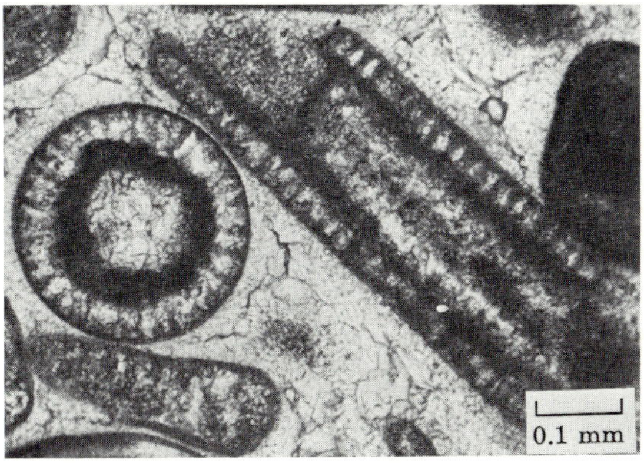

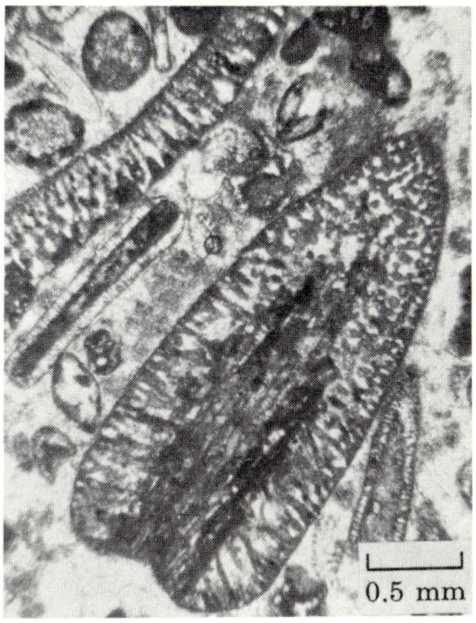

Fig. 89. *Ovulites*. Transverse and longitudinal sections of segments. Paleocene, Libya. Transmitted light, thin section.

Fig. 90. *Halimeda*. Longitudinal sections. Paleocene, Libya. Transmitted light, thin section.

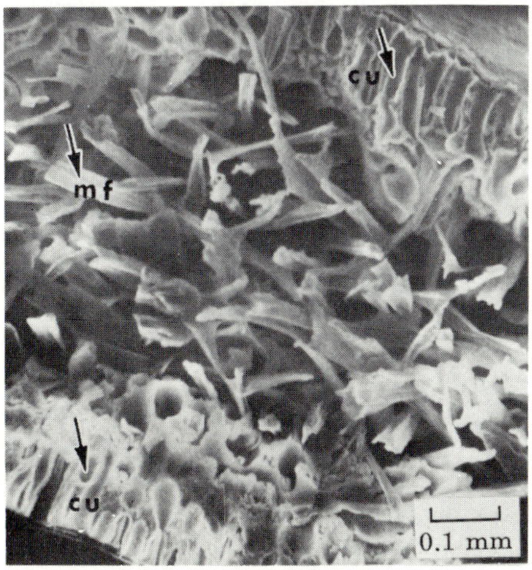

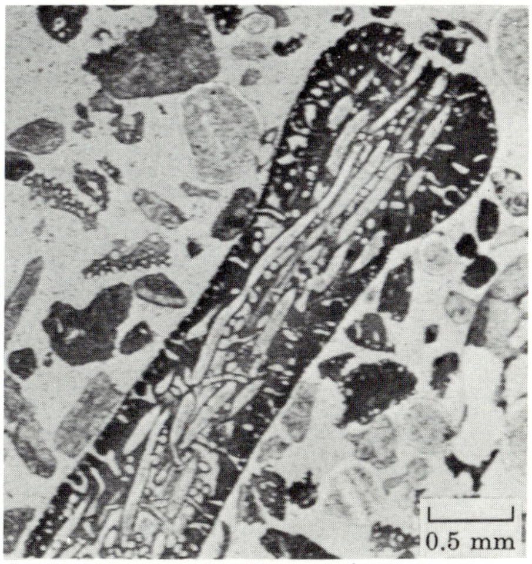

Fig. 91. *Halimeda*. Transverse section of segment showing medullary filaments (mf) and cortical utricles (cu). Recent, Florida. Scanning electron micrograph.

Fig. 92. *Halimeda*. Transverse section of segment. Recent, Florida. Transmitted light, thin section.

concluded that it is an organism probably related to colonial bacteria. He cautioned, however, that the origin of all "Microcodium" is not yet solved.

Nuia

Nuia consists of straight or curved calcareous tubes (averaging between 0.1-0.5 mm in diameter) with a dark, central canal. The wall is composed of radially arranged, calcite prisms (Fig. 94).

This genus has been assigned to both the calcareous green and blue-green algae. Maslov (1956) suggested a similarity between *Nuia* and *Microcodium* and grouped both in the same family (Microcodiaceae) as siphonaceous algae (Codiaceae), but later Maslov et al. (1963) assigned these genera to "problematical algae". Although *Nuia* superficially resembles the shape of primitive calcareous codiaceans, it lacks diagnostic features. Toomey and Klement (1966) carefully considered the systematic position of *Nuia* and concluded that it is best treated as a microorganism, possibly of algal affinities.

Nuia seems to be restricted to Lower Ordovician rocks in North America and the U.S.S.R., where it can be an important skeletal constituent in carbonate mounds.

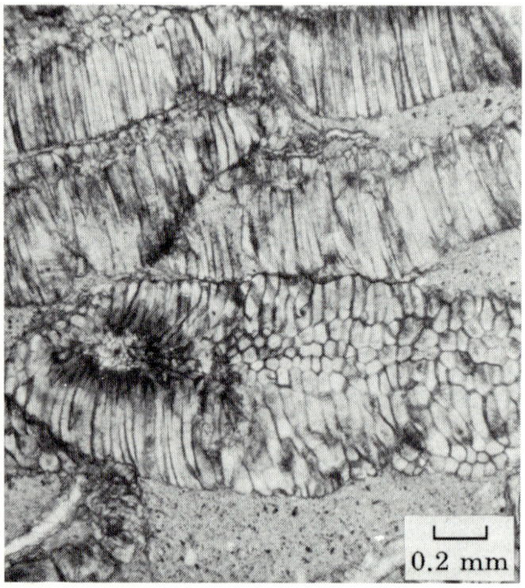

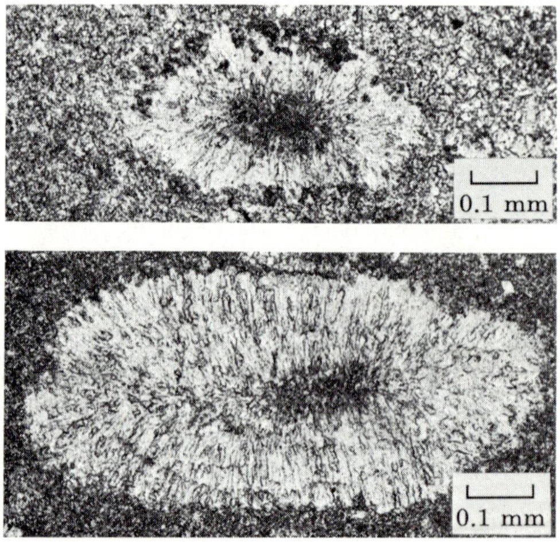

Fig. 93. *Microcodium*. Eocene, France. Transmitted light, thin section.

Fig. 94. *Nuia*. Transverse (above) and oblique section (below). Ordovician, Texas. Transmitted light, thin section. (Courtesy of D. F. Toomey).

Geologic range

Calcareous Codiaceae range throughout the Phanerozoic, except for the early part of the Cambrian (Fig. 95). They make up an important minor to major element in any given assemblage of warm-water, shallow-marine calcareous algae. Acmes of diversity and abundance occurred first in the Late Carboniferous and Permian (bladed and segmented forms), and again in the Late Mesozoic and Cenozoic (*Ovulites*, *Halimeda* and related genera).

Environmental distribution

The present-day distribution of living calcareous codiaceans provides a good basis for interpreting the environmental regimes of similar fossil forms (Table VI). *Halimeda* and related genera are exclusively marine algae restricted to tropical waters (25°C+ isotherm), except for one or two species of *Halimeda* which are known from subtropical regions.

Most forms colonize sand and mud substrates where rhizoids of the plant penetrate the soft bottom to develop holdfasts; however, Hillis (1959) noted exceptions, namely a few species of *Halimeda* attached to firm objects. Because of the delicate construction of these plants, they generally live below intense wave agitation.

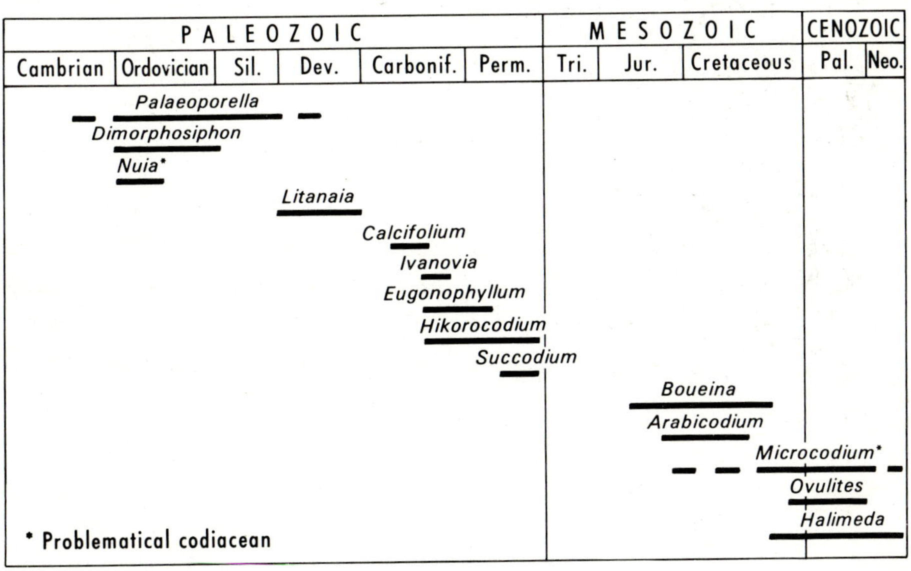

Fig. 95. Geologic ranges of representative genera of calcareous Codiaceae.

Light requirements and other factors permit these plants to range in depth from just below low tide to at least 100 m. They are most common and diverse at shallow depths of a few meters, especially in tropical marine shelf and lagoonal environments.

TABLE VI

FACTORS CONTROLLING DISTRIBUTION OF LIVING CALCAREOUS CODIACEAE

TEMPERATURE	Tropical (25°C+ isotherm); few species in subtropical waters.
DEPTH	Below low tide and extending to about 100 m; commonly a few meters.
SALINITY	Normal marine.
SUBSTRATE	Sand and mud; a few species attached to firm objects.
ENERGY	Low energy; below intense wave agitation.

Observations from submersibles in the deep fore-reef zones off Belize and Jamaica in the Caribbean (Wray, 1972a; Lang, 1974; and Moore et al., 1976) confirm that the distribution of *Halimeda* species is highly depth-dependent. *H. opuntia* is restricted to shallow water, several species are abundant in the range above 75 m, while *H. cryptica* is a deep-water form extending to approximately 100 m.

Although *Halimeda* is generally regarded as a shallow-water alga, its skeletal remains may be transported down steep slopes and accumulate in substantial amounts in bathyal environments. Also, the locally abundant growth of some species on fore-reef slopes is a significant deep source of sand-sized calcareous sediments for depositional environments below. Emery et al. (1954) noted a similar distribution of *Halimeda* plates in the Pacific atolls where they occur abundantly on floors of lagoons, as well as in sediments on the outer slope recovered in dredge hauls from over 600 m.

DASYCLADACEAE

This family was classified by Levring (1969) as one of two families belonging to the order Siphonocladiales. More often it is considered a single family of the order Dasycladales by paleontologists (e.g., Johnson, 1961) and botanists (Dawson, 1961; Bold, 1967); however, Valet (1969) believes the order consists of two families, Dasycladaceae and Acetabulariaceae (Table VII).

Characteristics

Living Dasycladaceae are represented by about eight genera, yet over 100 fossil genera, some dating back to the Cambrian, have been assigned to this family. Extant genera are tropical and subtropical marine plants with definite calcification, but of variable extent.

Plants grow upright and are attached to the substrate by rhizoids. Characteristically, the thallus has a radial symmetry about a long central axis (stem cell) that bears one or more whorls of lateral branches (Fig. 96). Reproductive organs (sporangia) occur on branches or in the central stem. Aragonitic encrustations preserve the morphology of the central axis, branches, and reproductive organs (Fig. 6). In effect, fossils are calcareous molds in which the stem and branches appear as canals or pores. They range in size from about 1 mm to several centimeters. Because whole plants may disarticulate, they often turn up in sediments as individual segments and fragments.

In living forms, an unusual situation takes place whereby the vegetative thallus is uninucleate, but a multinucleate condition develops prior to reproduction. Other strictly biological and cytological attributes, such as plastids and chemical composition of the cell walls, characterize the family which are not evident in fossils. Nonetheless, the preserved structures are so distinctive, easily recognized, and occur with such consistency as to make the Dasycladaceae a homogeneous and clearly defined group.

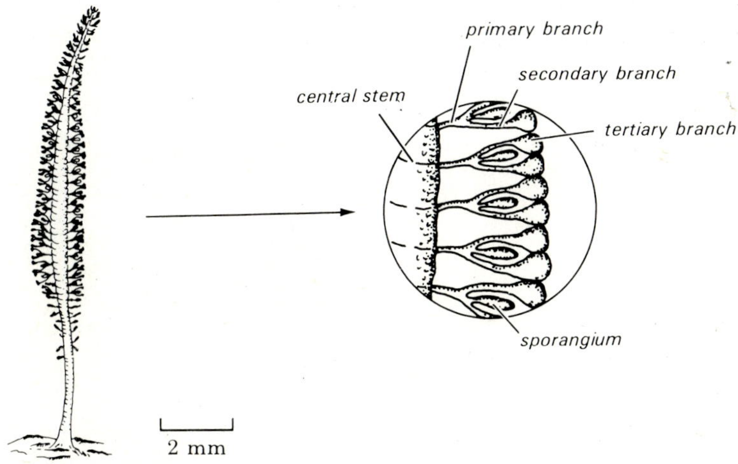

Fig. 96. Growth form and internal morphology of a living dasycladacean (*Neomeris*).

Classification

The dasycladacean morphology has produced a paradox in the systematic paleontology of the group. While distinctive features have been organized into seemingly logical classification schemes, they have at the same time been the source of differences of opinion among specialists. Pia (1920) proposed the first comprehensive classification system of genera and tribes, and this scheme has been widely used, or adopted with variations, by many paleontologists. For example, Kamptner (1958), Rezak (1959), Johnson (1961), Endo (1961), Herak (1965), Elliott (1968), and Kochansky-Devidé and Gusić (1971) have followed generally the work of Julius Pia in systematic treatments of the Dasycladaceae and their phylogenetic relationships.

Pia (1920) based his classification of fossil Dasycladaceae upon the following:

- General form of the thallus (cylindrical, club-shaped, etc.).
- Dimensions.
- Type of thallus or skeleton (segmented, unsegmented, etc.).
- Form of the whorled branches (simple or branched, broadened at extremities or tapered, open or closed pores, etc.).
- Arrangment of the whorled branches (irregular, in whorls, or in clusters).
- Shape of central stem or axial cell (cylindrical or other).
- Reproductive organs.

Although the primary branches may be arranged in a number of ways around the central axis, Pia (1920) described three patterns in fossil Dasycladaceae: an irregular arrangement (= aspondyle); arranged in regular whorls (= euspondyle); and arranged in whorls, but branches are regularly grouped in clusters (= metaspondyle) (Fig. 97).

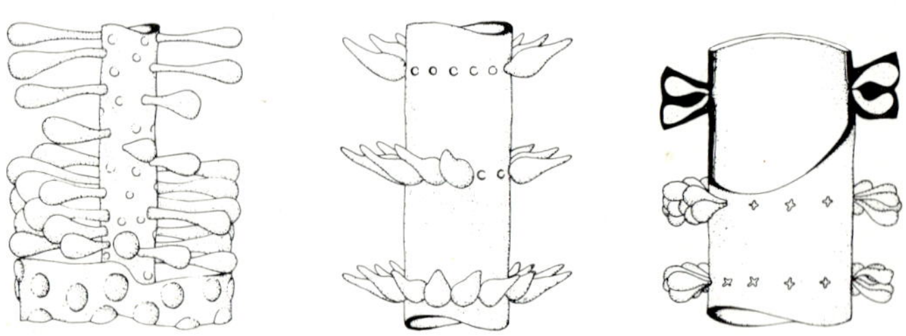

Fig. 97. Types of branching arrangements in fossil Dasycladaceae recognized by Pia. Left to right: aspondyle, euspondyle, and metaspondyle.

Reproductive organs may be developed in one of three positions: within the central stem (most Paleozoic genera are this type), on the primary branches, or within specialized gamentangia adjacent to secondary and tertiary branches.

In addition to these qualitative criteria, Pia (1920) introduced a system of letters designating various measurable parameters: D = exterior diameter; d = diameter of the axial stem; H = height of a segment; g = number of whorls per segment; h = distance between whorls; etc. Current authors use this system, although certain factors seem to have fallen into disuse and others have been added.

There is no consensus as to how many tribes belong to the Dasycladaceae. Originally, Pia (1920) proposed 11, but later he (1927) recognized 15. Johnson (1961) listed 16 tribes, but because a number of new tribes, mainly Paleozoic ones, have been erected in recent years, the total number is now greater. Until a modern monographic study of fossil dasycladaceans is undertaken, and certain taxonomic problems resolved, the number of valid genera and tribes will be difficult to ascertain.

Recently, Valet (1968, 1969) developed a revised classification of living Dasycladaceae from comprehensive studies of their morphology, cytology, and reproduction. Accordingly, extant genera are characterized by:

- Character of cortex formed by whorled branches.
- Type of thallus (segmented or unsegmented).
- Reproductive organs (single or clustered).
- Type of whorled branches and number of whorls.
- Type of reproductive function (direct zoospores or sporangia).

TABLE VII

CLASSIFICATION OF LIVING DASYCLADALES (VALET, 1969).

ORDER	*FAMILY*	*SUBFAMILY*	*TRIBE*	*GENUS*
Dasycladales	Dasycladaceae	Dasycladoideae	Dasycladeae	Dasycladus
				Chlorocladus
			Batophoreae	Batophora
		Bornetelloideae		Bornetella
		Neomeridoideae	Neomerideae	Neomeris
			Cymopolieae	Cymopolia
	Acetabulariaceae		Halicoryneae	Halicoryne
			Acetabularieae	Acetabularia

Species are based on other criteria, including the size and shape of first-order branches, presence or absence of second-order branches, and location of reproductive organs.

We see immediately differences between the classifications of Pia and Valet. Valet makes no mention of the general shape of the thallus in generic considerations, nor do dimensional criteria enter into the classification of living genera, although they define species.

The latest attempt to harmonize botanical and paleontological approaches to classifying dasycladacean algae is a critical review by a group of French paleontologists (Bassoulet et al., 1975). One of their principal conclusions is that criteria should be organized into a hierarchy of importance. They propose that the principal characteristic for designating genera (and tribes) is the arrangement of the primary branches, followed, in order, by the shape of the thallus, the shape and order of the branches, and reproductive organs. Dimensions, nature and degree of calcification, and the morphology of the axial stem should be utilized to characterize species.

Despite the apparent chaos in the classification of Dasycladaceae, the morphologies of many fossil genera are definitive, easily recognized, and separable from similar forms. Whether they conform to a biological classification is questionable. Obviously, any serious student of these fossils must confront the vast literature in the field; it is far too extensive and complex to summarize here.

Representative genera

A complete inventory of all, or even most, fossil dasycladacean genera and tribes is beyond the scope of this book. Bassoulet et al. (1975) estimate that more than 120 fossil genera have been described. The genera discussed in this section may be considered characteristic of the more common kinds of Dasycladaceae, and are generally arranged chronologically beginning with the earliest forms.

Amgaella

This genus has an unsegmented, cylindrical thallus (Fig. 98). Only primary branches are present, which are arranged in a spiral pattern. Reproductive organs are unknown, probably because they developed within the central stem and were not preserved. *Amgaella* is similar to *Siberiella*, *Vologdinella*, and other genera which have been described from the Early and Middle Cambrian of the U.S.S.R. These and other early dasycladacean genera have been assigned to several tribes by Korde (1961).

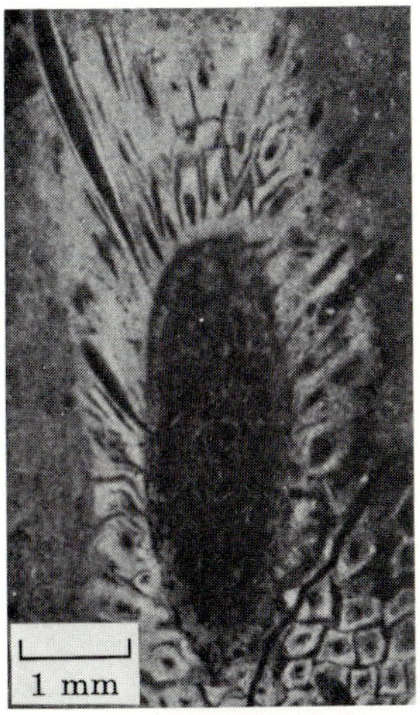

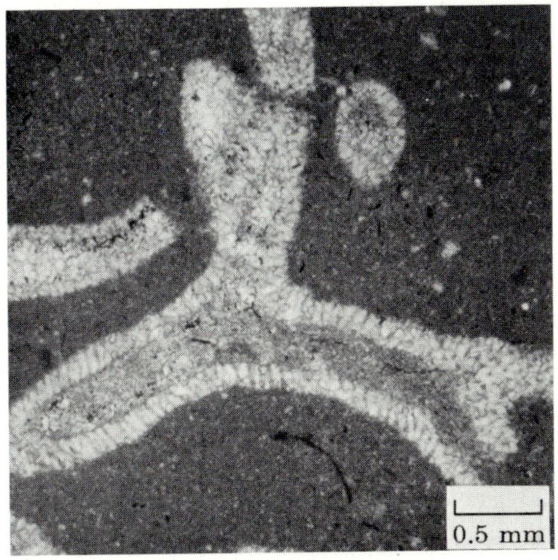

Fig. 98. *Amgaella*. Cambrian, U.S.S.R. Transmitted light, thin section. (From Korde, 1961).

Fig. 99. *Vermiporella*. Ordovician, Norway. Transmitted light, thin section.

Rhabdoporella and Vermiporella

These genera are characterized by irregularly arranged (aspondylic), primary branches (Fig. 97). *Rhabdoporella* is essentially a straight cylinder with a large central stem, whereas *Vermiporella* is distinguished by a bifurcated thallus (Figs. 99, 100). Sporangia are unknown, but presumably were borne in the central stem. Both genera have been considered among the most primitive Dasycladaceae (Johnson, 1961), but Kozlowski and Kazmierczak (1968) found silicified specimens of *Vermiporella* that suggest a much more advanced morphology (three orders of euspondylic branches).

Cyclocrinites

This dasycladacean is unsegmented, and although the shape varies, it is distinctly inflated or globular (Fig. 101). Primary branches are long and slender and they terminate in clusters of secondary branches, which form a faceted pattern of cortical depressions on the outer surface. The morphology and distribution of *Cyclocrinites* have been described in detail by Nitecki (1970).

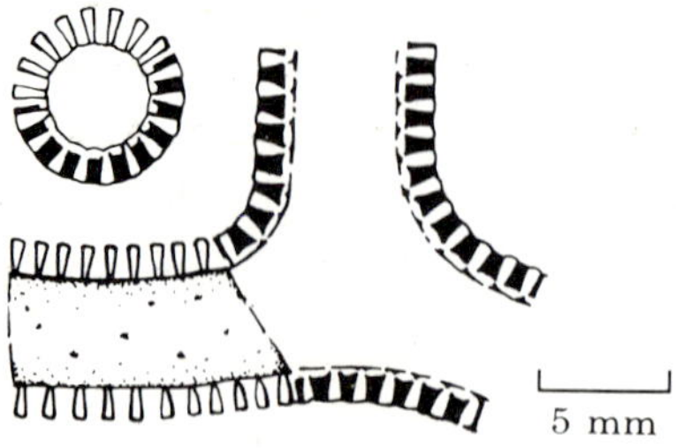

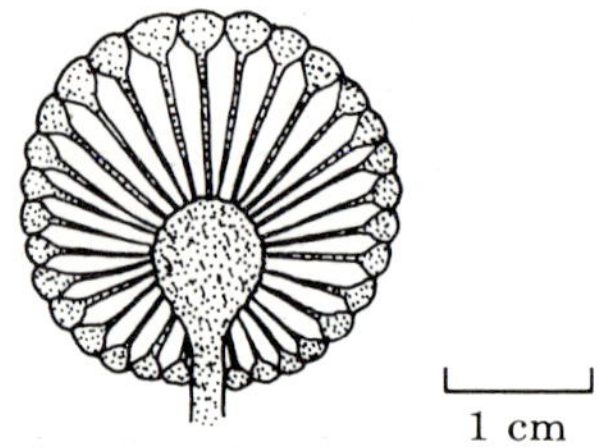

Fig. 100. *Vermiporella*. Diagrammatic sections of bifurcated thallus. (After Pia, 1920).

Fig. 101. *Cyclocrinites*. Diagrammatic section of entire thallus.

Epimastopora

Epimastopora presumably had an unusually large, cylindrical, unsegmented thallus, but the exact size and shape are speculative because it is known only from fragments (Fig. 102). Primary branches occur in regular, closely spaced whorls. Sporangia are unknown, but probably occurred in the axial stem.

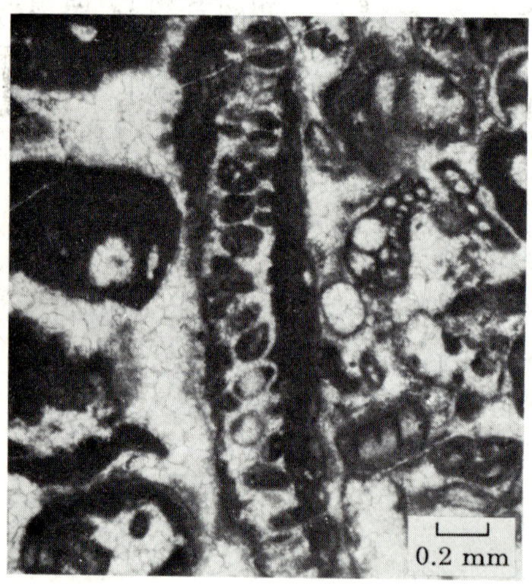

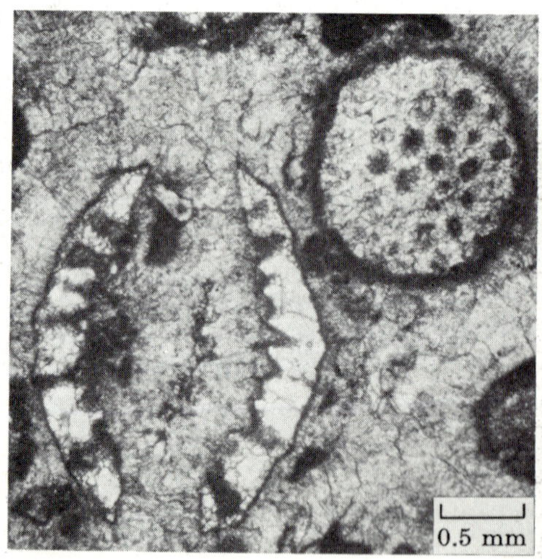

Fig. 102. *Epimastopora*. Upper Carboniferous (Pennsylvanian), Colorado. Transmitted light, thin section.

Fig. 103. *Mizzia*. Permian, Texas. Transmitted light, thin section.

Mizzia

Mizzia is composed of spherical or cylindrical segments, which presumably articulated (Fig. 103). The central stem is barrel-shaped with constrictions at the joints between segments. Only primary branches are present; these are arranged in concentric rows and they enlarge in diameter abruptly near the exterior. The termination of the branches produces a hexagonal pattern on the surfaces of segments. *Cyclocrinites*, *Epimastopora*, and *Mizzia*, among others, have all been classified within the tribe Cycloc riniteae (Johnson, 1961).

Beresella, Dvinella and Kamaena

These genera are a few of several similar Late Paleozoic tubiform fossils that have been variously considered to belong to different algal groups and foraminifera. They are often assigned to the Dasycladaceae as member of several tribes, including the Beresselleae (Maslov and Kulik, 1956; Kulik, 1964), although Riding and Jansa (1974) exclude one genus (*Uraloporella*) from the Dasycladaceae and question the algal affinity of others.

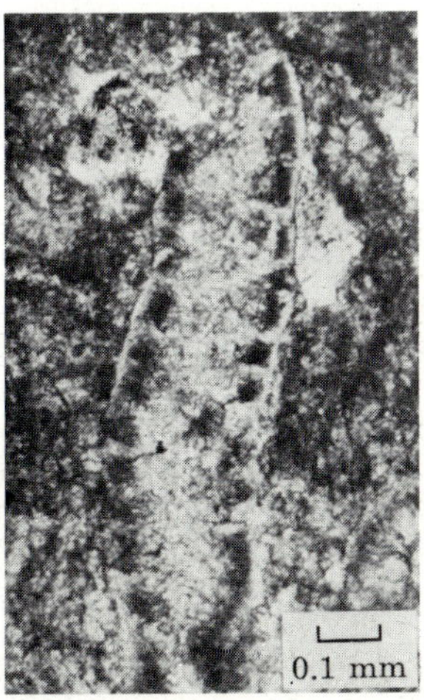

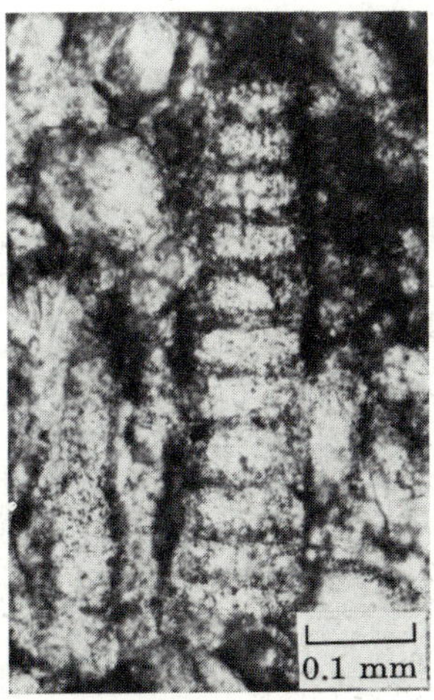

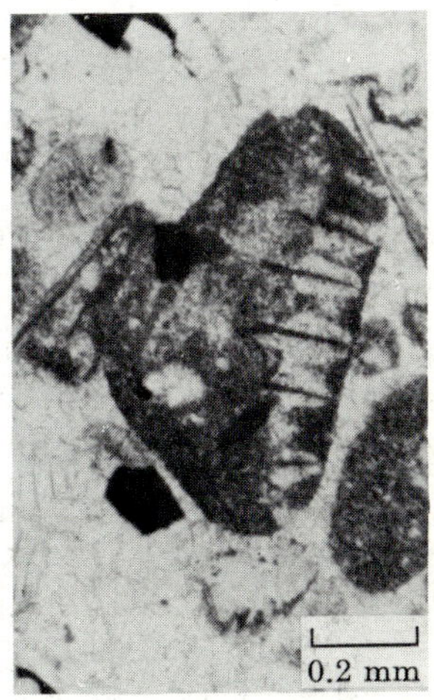

Fig. 104. *Dvinella*. Upper Carboniferous (Pennsylvanian), Nevada. Transmitted light, thin section.

Fig. 105. *Kamaena*. Devonian, Alberta. Transmitted light, thin section.

Fig. 106. *Koninckopora*. Lower Carboniferous, Japan. Transmitted light, thin section.

Inarticulate, bifurcated, cylindrical thalli characterize these genera (Figs. 104, 105). Some forms have distinctive horizontal partitions which nearly close the axial stem and give the appearance of internally segmented tubes. Three orders of branching have been recognized in the walls of the thallus (Elliott, 1970), but more commonly only primary branches are preserved. These fossils are abundant and geographically widespread in Devonian and Carboniferous limestones (Rich, 1967; Petryk and Mamet, 1972; and Mamet and Roux, 1974).

Koninckopora

Known only from fragments, this genus had an elongate, cylindrical thallus with a large central stem. According to Johnson and Konishi (1956) some plants may have been 50 cm long and more than 4 cm in diameter. *Koninckopora* has closely spaced, short, cylindrical or subpolygonal branches which have a distinctive pattern in section (Fig. 106). This Carboniferous genus seems to be restricted to rocks of the Visean stage.

Diplopora

The thallus is cylindrical and club-shaped with a central stem of variable diameter (Fig. 107). Primary branches usually occur in clusters of three or six; these vary in shape and length among the different species. Sporangia were developed in the central stem. *Diplopora* is a common dasycladacean in Triassic limestones of central Europe.

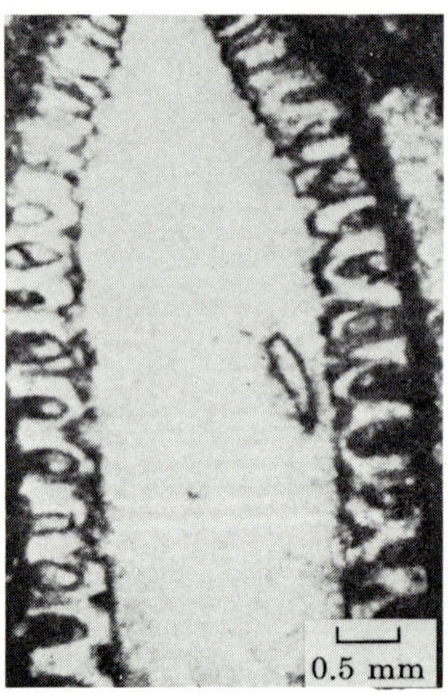

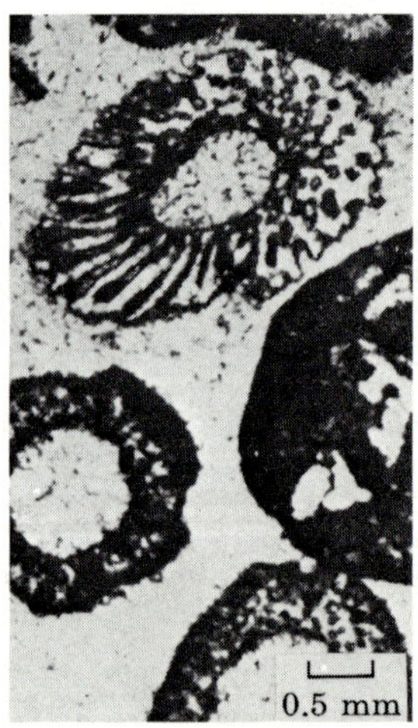

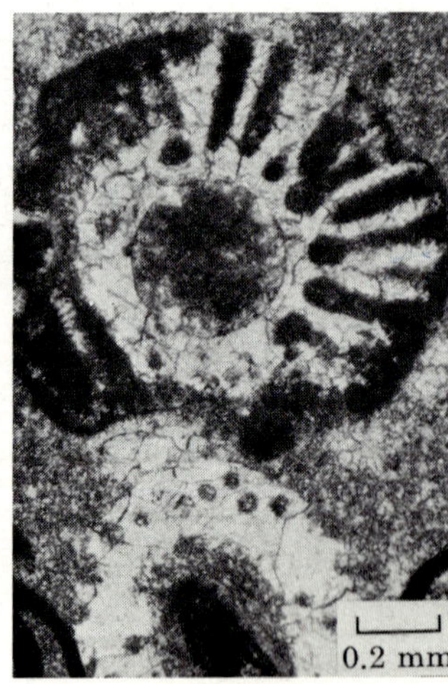

Fig. 107. *Diplopora*. Triassic, Yugoslavia. Transmitted light, thin section. (From Herak, 1965).

Fig. 108. *Macroporella*. Triassic, Yugoslavia. Transmitted light, thin section. (From Herak, 1965).

Fig. 109. *Clypeina*. Paleocene, Libya. Transmitted light, thin section.

Macroporella

A cylindrical thallus with a rather large central stem characterizes *Macroporella* (Fig. 108). Robust primary branches thicken towards the exterior and are arranged in irregular whorls. Secondary branches are absent, and sporangia presumably developed in the central stem.

Clypeina

In life the cylindrical thallus of *Clypeina* appeared to alternately pinch and swell due to widely spaced whorls of primary branches (Fig. 110). As a result of the structure and calcification, the plants disaggregated into bowl-shaped discs corresponding to the whorls of primary branches (Fig. 109). The central stem is moderately large and sporangia occur in the lower ends of the branches.

Cylindroporella

The thallus is segmented, consisting of cylindrical segments with rounded ends (Johnson, 1954). Six short primary branches developed in numerous whorls from a relatively small central stem. Sporangia are large and occur on the lower ends of the primary branches (Fig. 112). In vertical sections, the sporangia and branches appear to alternate in position. Secondary branches occur in clusters.

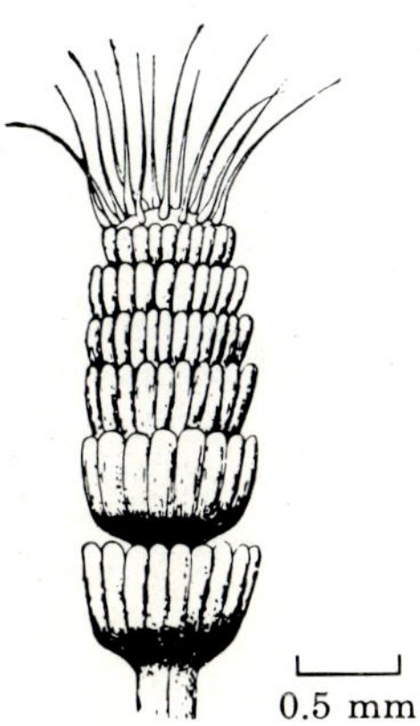

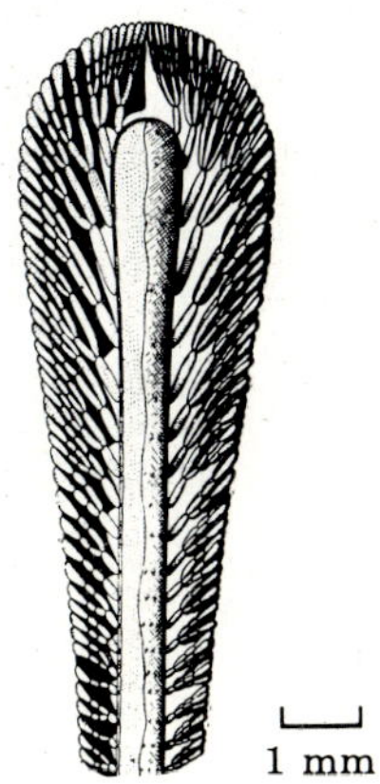

Fig. 110. Reconstruction of *Clypeina*. (After Morellet, 1918).

Fig. 111. Reconstruction of *Palaeodasycladus*. (After Pia, 1920).

Trinocladus

The cylindrical thallus of *Trinocladus* has a moderately large central stem. Primary branches, occurring in regular whorls, give rise to secondary branches, and these in turn to clusters of tertiary branches. Lower whorls may show only primaries (Fig. 113), whereas upper parts of the plant preserve the full range of branch forms. Primary branches are thick, widen outward, and probably contain the sporangia.

Palaeodasycladus

This fossil dasycladacean has an elongate, cylindrical (club-shaped) thallus consisting of numerous, regular whorls of primary branches which are inclined to the central stem. Pia's reconstruction of the plant (Fig. 111) shows the well developed nature of primary, secondary, and tertiary branches. The latter two types are subdivided into clusters of four to six branches. All branches are slightly inflated. Sporangia are unknown.

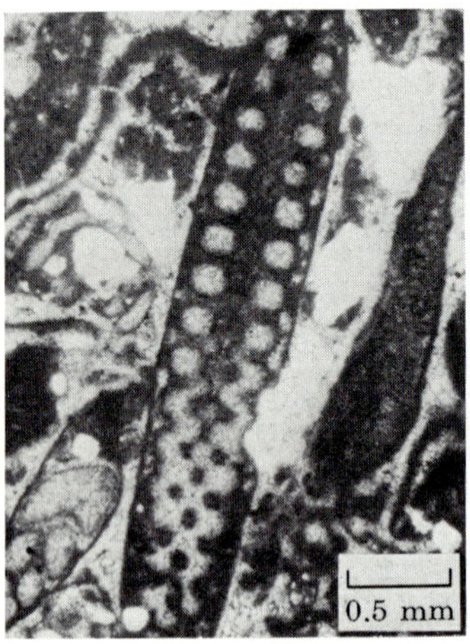

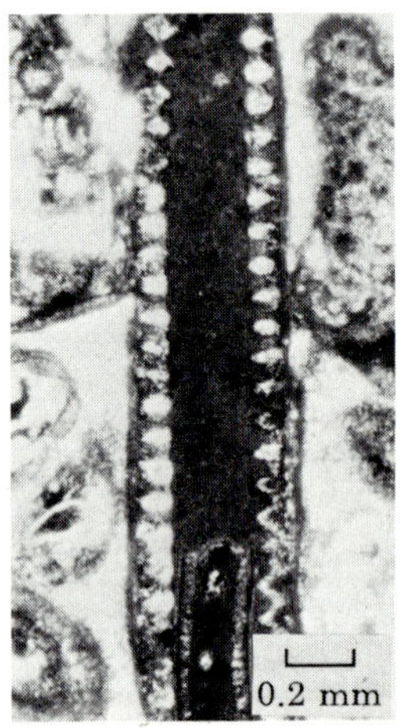

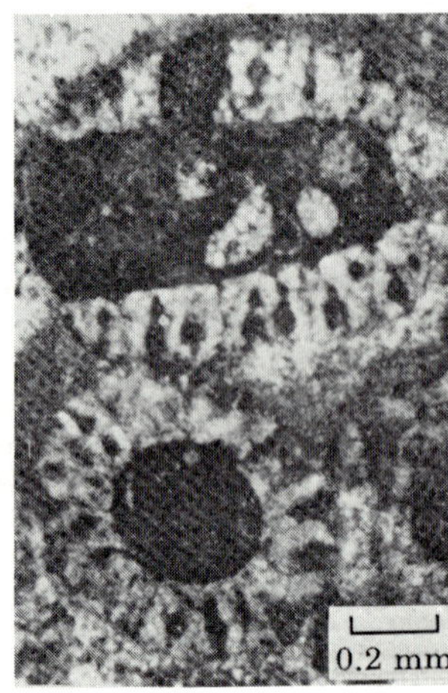

Fig. 112. *Cylindroporella*. Cretaceous, Texas. Transmitted light, thin section.

Fig. 113. *Trinocladus*. Section showing only primary branches. Paleocene, Libya. Transmitted light, thin section.

Fig. 114. *Neomeris*. Cretaceous, Texas. Transmitted light, thin section. (From Johnson, 1968).

Neomeris

Neomeris is an extant genus distinguished by a small (1-2 cm high), unsegmented, cylindrical thallus (Fig. 96). The central stem bears regular whorls of primary branches, and characteristically each primary branch divides into a stalked sporangium and two secondary branches set in the same plane (Fig. 114). Calcification surrounds the sporangia and secondary branches, but the thallus is essentially uncalcified around the central stem and primary branches. The genus was founded on Recent species, and so there are differences of opinion as to the concept of some fossil forms, especially in the Early Cenozoic and Cretaceous.

Cymopolia

Cymopolia has a dichotomously branched, segmented thallus, which may be as much as 25 cm high (Fig. 115). Individual segments are cylindrical with a relatively large central stem from which develop regular whorls of primary branches (Fig. 116). These primaries divide into usually four secondaries and one sporangial cavity. The secondary branches reach the outer surface to form a dense pore-pattern. The plant invariably disarticulates into individual segments after death.

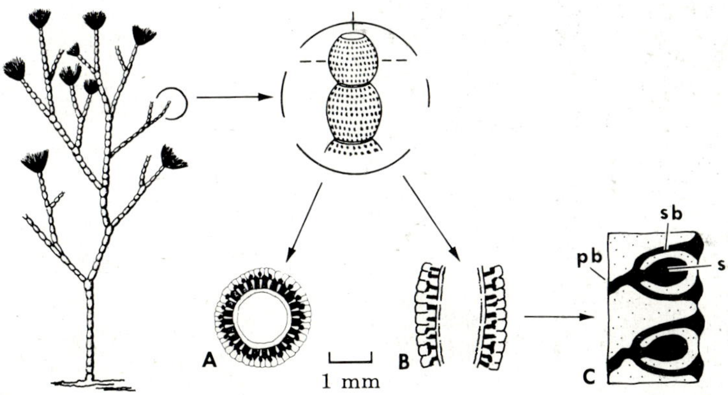

Fig. 115. Segmented dasycladacean alga (*Cymopolia*). Segments illustrating central axis and lateral branches shown diagrammatically. A) Transverse section. B) Longitudinal section. C) Detail of branches and encased sporangia. Primary branch (pb), secondary branch (sb), sporangium (s).

Acetabularia and Acicularia

These two extant genera are closely related, and *Acicularia* is often considered a form of *Acetabularia* by botanists, although fossils of the two genera have different geologic ranges. These delicate and distinctive plants, a few centimeters high, have a slender central stem bearing an inverted umbrella-shaped disc at the apex (Fig. 118). This specialized calcareous disc is made up of radially arranged rays that contain spherical sporangial cavities. Fossils are usually individual calcareous rays or fragments resulting from the disaggregation of the apical disc (Fig. 117).

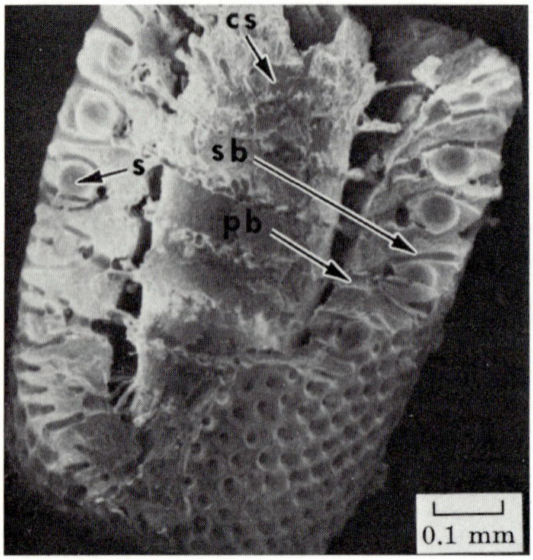

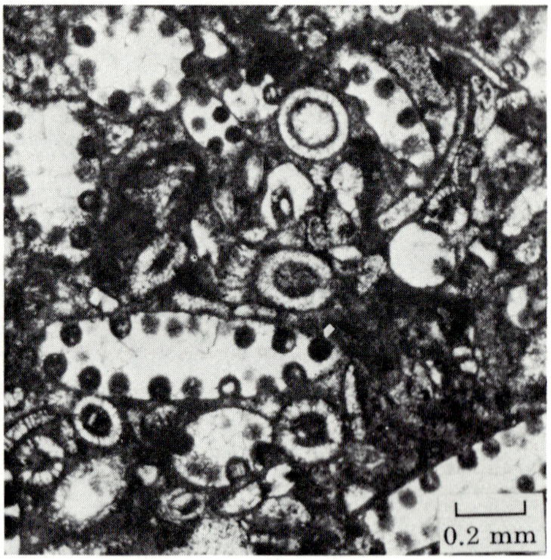

Fig. 116. *Cymopolia*. Structural detail of segment. Central stem (cs), primary branch (pb), secondary branch (sb) and sporangial cavity (s). Recent, Florida. Scanning electron micrograph.

Fig. 117. *Acicularia*. Limestone composed dominantly of fragments of this alga. Paleocene, Libya. Transmitted light, thin section.

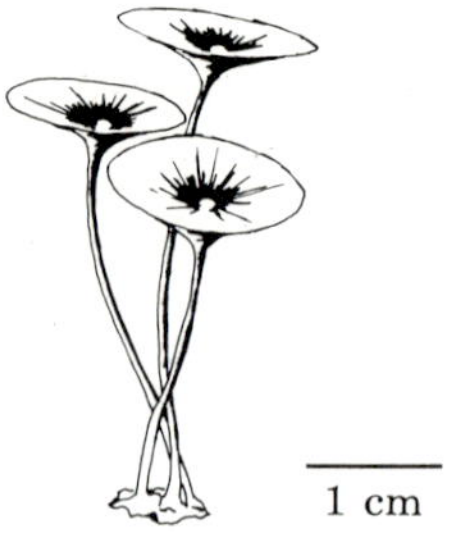

Fig. 118. Growth habit of *Acetabularia*.

Receptaculitids -- dasycladacean algae ?

The receptaculitids generally have been described as sponges or problematical fossils. Recently, several authors (e.g., Kesling and Graham, 1962; Byrnes, 1968; Rietschel, 1969; and Nitecki, 1970, 1972) have concluded that most taxa are calcareous green algae because their morphology shows close affinities to the plant kingdom. Nitecki (1972) assigned these spongelike fossils to a family within the Dasycladales, in which he also included the previously recognized dasycladacean tribe Cyclocriniteae, while Rietschel (1969) considers the receptaculitids a separate order of green algae. Meanwhile, others continue to support an animal affinity for the group.

Receptaculitids are usually globular in shape, ranging from a few centimeters to about 30 cm in diameter. The internal morphology consists of a central axis and radially arranged branches which appear as hexagonal or rhombohedral facets on the surface of the fossil (Fig. 119).

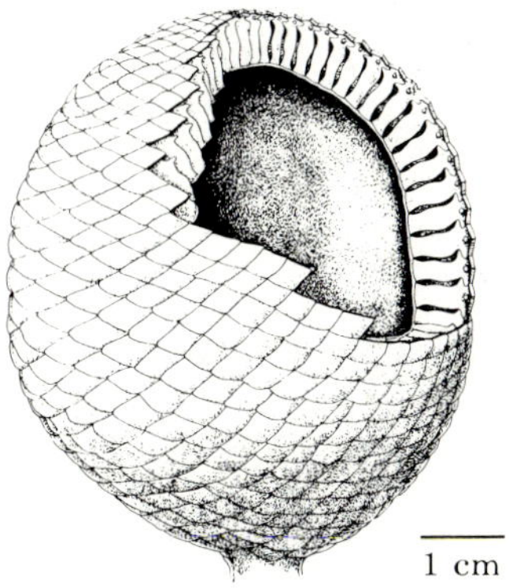

Fig. 119. Reconstruction of *Receptaculites sacculus*. (From Nitecki, 1972).

Receptaculitids are common in the Ordovician and Silurian, but range into the Permian. They are widespread geographically, occurring on all modern continents except Antarctica. The group is predominantly associated with carbonate facies, including reefs, and is believed to be indicative of warm-water, marine environments (Nitecki, 1972a).

Calcispheres

Calcareous, spherical bodies, commonly called calcispheres, are widely known from Upper Paleozoic limestones, especially Devonian and Carboniferous, but also occur in older and younger strata. All are hollow, have well defined walls (often composed of several layers), and usually measure 75-200 microns in diameter. In addition, some may have radially arranged spines on the outer surface (Fig. 120).

The origin of calcispheres is a matter of debate and they have been referred to various plant and animal groups. Stanton (1963) suggested they represent some form of plant spore or reproductive body. Later, Rupp (1967) noted the similarity of nonornamented fossil calcispheres and reproductive cysts of living dasycladacean algae belonging to the subfamily Acetabulariae. These algae develop calcareous reproductive bodies closely resembling the size and wall structure of the fossil calcispheres (Fig. 121). Recently, Marszalek (1975) described the calcisphere-producing alga *Acetabularia* in south Florida.

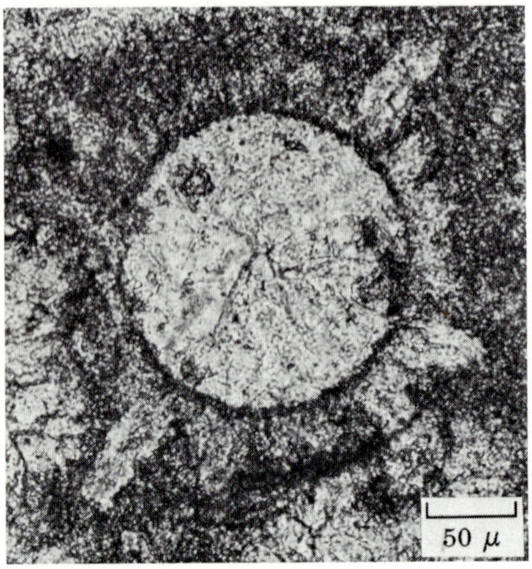

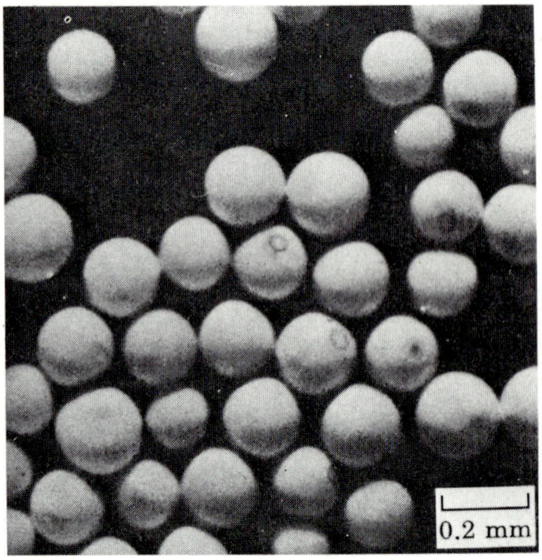

Fig. 120. Calcispheres. Devonian, Western Australia. Transmitted light, thin section.

Fig. 121. Calcareous reproductory cysts from *Acetabularia antillana*. (From Rezak, 1971).

Fossil calcispheres occur in rocks interpreted as having formed in shallow-water, restricted-circulation environments. This parallels the habitat of living Acetabularieae. It seems very probable that many fossil calcispheres owe their origin to dasycladacean algae, and as such, they provide a useful paleoenvironmental indicator.

Geologic range

The Dasycladaceae have a long and continuous chronologic range extending from the Early Cambrian (Fig. 122). Some rather simple morphologies characterize the Early Paleozoic flora; however, if the receptaculitids belong to the family, or are closely related, they represent an extraordinary development in size and complexity at an early stage in the evolution of these algae. Several living genera can be traced back in time to the Early Cenozoic and Cretaceous.

Relatively common and diverse assemblages occur in the Ordovician and in the Carboniferous and Permian. However, the large number of genera and species described from the Mesozoic suggests that the family reached an acme of development during this time.

Dasycladacean species have potential value in biostratigraphic zonation in some regions. Elliott (1968) found that many species were restricted to stage-level ranges in the Mesozoic of the Tethyan seaway.

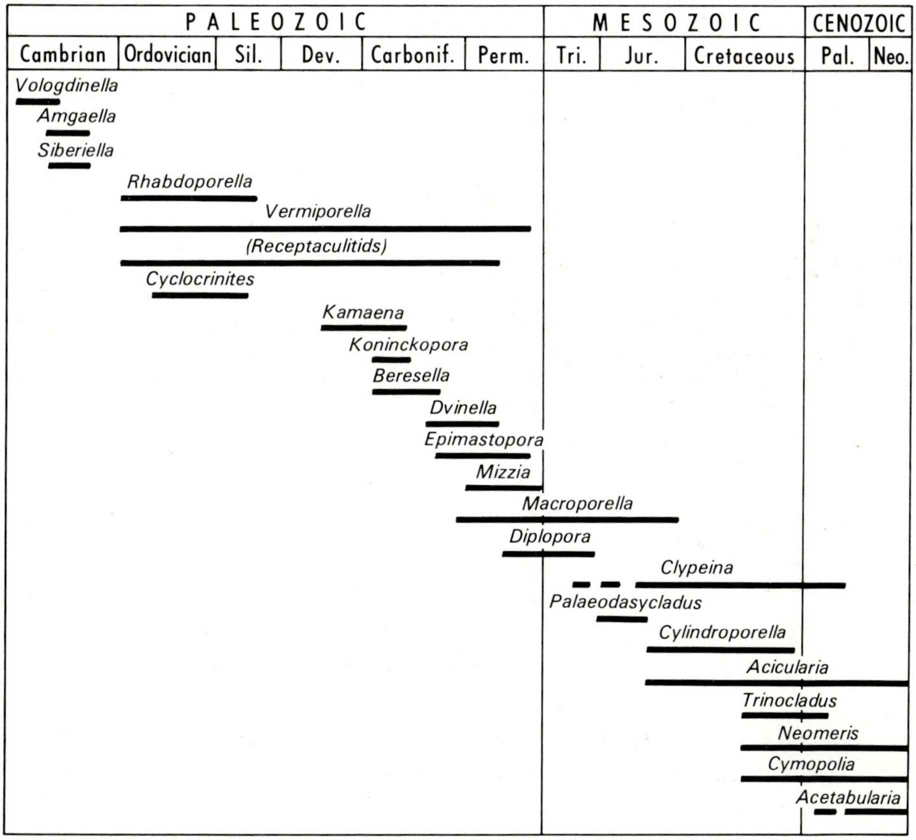

Fig. 122. Geologic ranges of representative Dasycladaceae.

Environmental distribution

Dasycladacean algae have a distribution similar to calcareous codiacean genera; that is, they inhabit mainly shallow, warm, marine waters (Table VIII). However, there are significant differences between the two groups. First, living dasycladacean algae are relatively inconspicuous in the total marine algal flora, and much less evident than the calcareous codiaceans. This may seem anomalous when one considers the vast number of fossil dasycladacean versus the few codiacean genera, but of course, many codiaceans have disaggregated into unidentifiable grains. Second, there are minor but important differences between the two groups in their tolerances to temperature, salinity, and depth.

TABLE VIII

FACTORS CONTROLLING DISTRIBUTION OF LIVING DASYCLADACEAN CALCAREOUS ALGAE

TEMPERATURE	Tropical and subtropical; few species in warm temperate waters.
DEPTH	Below low tide to about 30 m; commonly less than 5 m.
SALINITY	Mostly normal marine; some species tolerate hypersaline to brackish waters.
SUBSTRATE	Sand and mud; some species attached to firm objects.
ENERGY	Low energy; below intense wave agitation or protected areas.

Living dasycladaceans occur most often in tropical and subtropical marine waters, but a few inhabit warm-temperate environments. As a group they seem to prefer normal marine salinities, but the Acetabularieae, for example, tolerate salinities ranging from hypersaline to brackish, and wide variations in temperature.

The depth range of dasycladacean algae normally extends from just below low tide to about 30 m, but Elliott (1968) cited an occurrence down to 90 m in the eastern Mediterranean. Living *Neomeris* do not exceed depths of 10 m, and most species live at less than 5 m (Konishi and Epis, 1962). The maximum abundance of living dasycladaceans would seem to be in a depth range down to about 5 m, but existing in diminishing numbers down to 10 m with rare occurrences to several tens of meters.

Most dasycladaceans colonize sand and mud bottoms, but many forms attach their rhizoids to firm objects, such as shells, in these loose-sediment environments. They are found only in low-energy regimes, either below wave base or in protected situations, and are a typical algal flora in marine lagoons. Based on extensive studies of Permian to Early Cenozoic dasycladacean assemblages and associated biota of the Middle East, Elliott (1968) concluded that the ecological requirements of fossil dasycladaceans were essentially the same as those of living descendants.

CHAROPHYCEAE

Members of this class are physiologically and biochemically similar to other green algae and are classified in the Chlorophyta by Levring (1969); however, they have a very different morphology, as well as reproduction, and often have been assigned to a separate phylum (Grambast, 1974). They possess a number of morphological and cytological characters of higher plants, and may occupy an isolated position between the green algae and the bryophytes.

Geologically, charophytes are treated separately from other calcareous algae because they are mainly freshwater plants whose calcified remains are used mainly for biostratigraphic zonation of nonmarine sediments. Thus these algae are not usually a concern of carbonate sedimentologists and paleontologists whose efforts are focused on marine sediments. Details of the external morphology are critical to identification, necessitating whole specimens separated from the rock matrix, although some specialists emphasize the need for thin sections in addition.

Characteristics

The charophyte plant attains a size of several centimeters to as much as a meter, and overall resembles the higher plants in its general form and external differentiation of parts (Fig. 123). They have an erect, branched thallus divided into a regular succession of nodes, which bear whorls of small branches, and internodes. The sex organs, both male (antheridium) and female (oogonium), develop from the nodes during fertile stages of the plant.

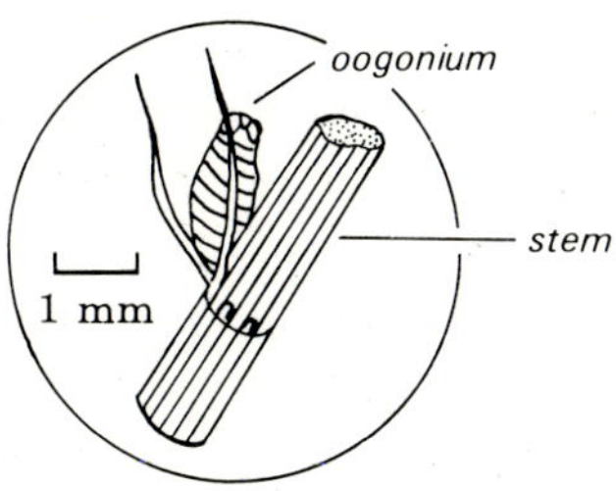

Fig. 123. Charophyte plant showing location of calcified oogonium (insert).

Most genera calcify only the reproductive organs, principally the oogonium. Antheridia on living charophytes do not usually calcify.

Characteristically, oogonia are elliptical in shape, 0.5-1.0 mm in size, and consist of various arrangements of tubular cells. The oogonial wall of species of the Characeae is composed of five enveloping cells that spiral to the left (Fig. 124A). On the distal ends of the enveloping cells (apex of the oogonium) of some species is the coronula, a crown-like structure made up of smaller cells. Other morphological variations include forms in which the enveloping cells are vertical rather than spiral (Sycidiaceae), (Fig. 124B), and dextrally spiraled forms (Trochiliscaceae) (Fig. 124C). The Clavatoraceae are Mesozoic charophytes in which the oogonia are enveloped in utricles (Fig. 124D). The Characeae, the only extant family, is differentiated from all other charophytes by the disappearance of the apical pore because of the close junction of the tips of the spiral cells at the apex.

The fossil record of the Charophyceae is based almost entirely on descriptions of gyrogonites, the spiral calcified parts of the oogonia. Fossil vegetative parts of these plants are essentially unknown.

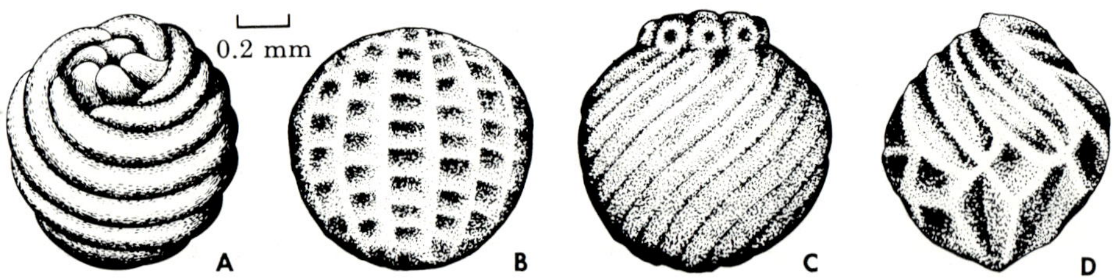

Fig. 124. Typical forms of fossil gyrogonites. A) Sinistrally spiraled form belonging to family Characeae. B) Gyrogonite of vertical enveloping cells (family Sycidiaceae). C) Dextrally spiraled form (family Trochiliscaceae). D) Gyrogonite enveloped in utricles (family Clavatoraceae).

Classification

This group of algae is represented by a larger number of fossil members than extant species. At least 50 genera and over 300 species of fossil charophytes have been described. Living forms belong to a single class (Charophyceae) consisting of one order (Charales) and one family (Characeae) with two tribes or subfamilies, the Chareae and Nitelleae, and about five genera. In North America, for example, living Characeae are represented by three genera and 22 species.

Grambast (1974) has recognized nine families of fossil charophytes (including the Characeae) all of which are assigned to the phylum Charophyta. Five families, Sycidiaceae, Chovanellaceae, Trochiliscaceae, Eocharaceae, and Palaeocharaceae are restricted to the Paleozoic and one, Clavatoraceae, to the Mesozoic; the other three families, Porocharaceae, Raskyellaceae, and Characeae, include primarily Mesozoic and Cenozoic taxa.

The classification of the charophytes has been influenced by many changing ideas about the relationship of these unique plants to other groups, as well as phylogenetic relationships within the group. Also, it seems that the taxonomic viewpoints of botanists and paleobotanists differ on some matters. Monographic studies of the group and other pertinent references include works by Horn af Rantzien (1959), Maslov (1963), and Wood and Imahori (1964, 1965).

Geologic range

Charophytes have a continuous record since the Late Silurian (Fig. 125), and the group has undergone numerous evolutionary changes through time which are recognizable in fossil populations of gyrogonites (Grambast, 1974). Only two genera are known from the Silurian, but by Devonian and Early Carboniferous time a high degree of morphological variations of gyrogonites is observed compared to Recent forms. Over 20 genera occur in the Early Cretaceous, the height of generic diversity, and the number declined to 10-12 through the Late Cretaceous and the Early Cenozoic.

Relatively rapid evolution, numerous species, and distinctive morphologies of the charophytes have combined to make these fossils an excellent group for developing biostratigraphy in nonmarine sediments, especially lacustrine deposits. To illustrate, the morphologically diversified Paleogene flora of Western Europe has been used to subdivide the Paleocene and Eocene into 12 zones and the Oligocene into five (Grambast, 1972). Using three evolutionary lineages of the family Clavatoraceae, Grambast (1974) recognized nine zones in the Early Cretaceous.

The present understanding of the biostratigraphic utility of the group has developed within the past two decades. Earlier, Peck (1957) recognized only 12 genera in the entire Mesozoic of North America, and he was able to delimit just a few taxa to stage levels. It seems clear now that modern zonation schemes based on charophytes are useful in correlating nonmarine sediments on a worldwide scale.

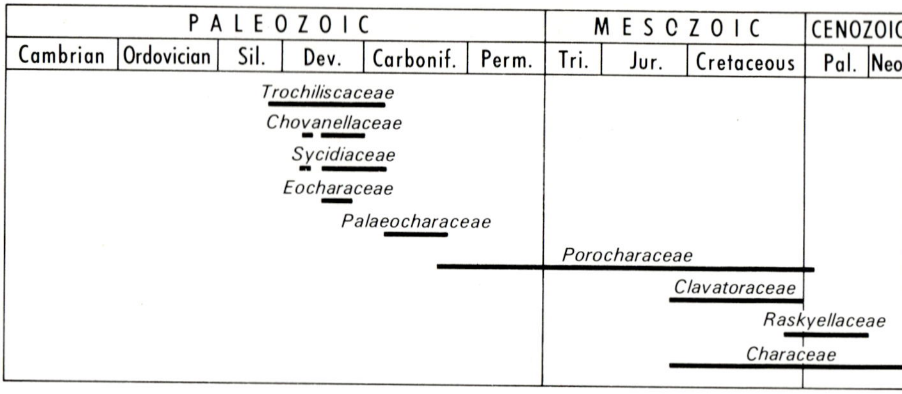

Fig 125. Geologic ranges of charophyte families. (After Grambast, 1974).

Environmental distribution

Living charophytes inhabit fresh and brackish-water environments, in contrast to the marine habitats of other calcareous green and red algae, and fossil forms presumably occupied similar nonmarine environments. Most species grow submerged in fresh standing water and upon a muddy or sandy bottom. They often form extensive subaquatic meadows that may extend to depths of at least 10 m. Calcified remains have accumulated in substantial amounts on ancient lake bottoms (Fig. 126), and recent accumulations are recovered as commercial sources of agricultural lime and cement.

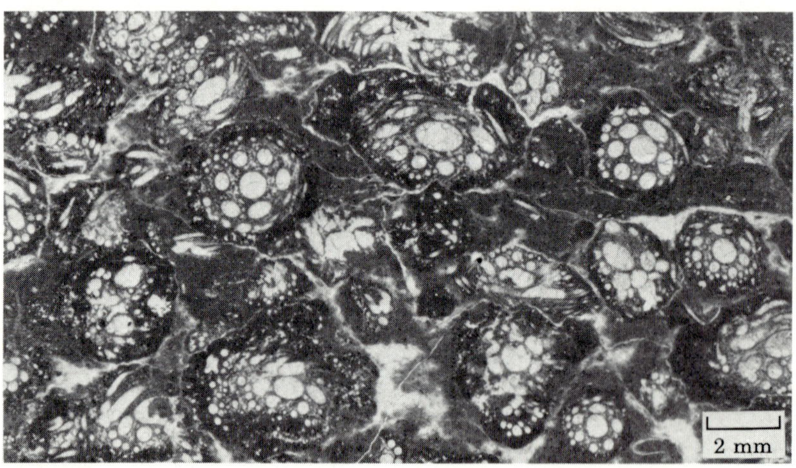

Fig. 126. Numerous sections of charophyte gyrogonites in freshwater limestone. Jurassic, Colorado. Transmitted light, thin section.

Most species of *Chara* thrive in clear, hard-water situations, whereas some *Nitella* prefer soft water. A few species of *Chara* and *Nitella* grow in brackish-water environments near sea coasts. Gyrogonites may be transported from their place of origin and deposited in marginal marine settings.

Chapter 6

ALGAL-LAMINATED SEDIMENTS AND STROMATOLITES

Nonskeletal algae have created a vast and fascinating assortment of calcareous structures dating back to early Precambrian time. These forms are biosedimentary in origin, resulting from the interaction of filamentous and coccoid algae and physical sedimentary processes. In effect, the algae have served a sediment-stabilizing role by mechanically trapping and binding sediment particles on organic films. The process has resulted in various kinds of products, ranging from well laminated stromatolites and oncolites to crudely laminated structures and clotted fabrics.

The study of algal-laminated sediments and stromatolites is a recent phenomenon; the majority of the investigations have been done since 1960. The broad scope of the subject today is emphasized by the nearly-800-page volume "Stromatolites" edited by Walter (1976). Interpretations of biosedimentary structures require integrated approaches, involving sedimentology, paleontology, and biology.

This chapter outlines the characteristics and distribution of stromatolites and other products of sediment-binding algae. Despite a considerable amount of attention by both paleontologists and sedimentologists, many problems and some misconceptions persist; consequently, it is important to keep in mind the following:

- An enormous variety of biogenic sedimentary structures and fabrics have been created by nonskeletal blue-green and green algae.
- Stromatolites are not exclusively intertidal and supratidal features, although they occur most often along margins of marine basins and saline lakes.
- Algal-laminated sediments and stromatolites are not always separable from inorganically produced structures with similar morphologies, such as caliche (calcrete).
- Stromatolites are rarely sufficient by themselves to solve a particular problem, and it is necessary to consider evidence of associated sedimentary structures and fossils.

THE ROLE OF ALGAE

Communities of nonskeletal blue-green algae forming organic films of mainly mucilaginous material may effectively stabilize sedimentary surfaces by trapping and binding clastic particles of sediment. This is the basic organism-sediment interaction that produces stromatolites and other algal-laminated sediments. When an algal film is covered by a layer of sediment, filaments penetrate the material and colonize the surface to reform a new algal film. The usual product of this process is a succession of laminae in which one type is sediment-rich and the other is organic-rich, made up mainly of blue-green algae (Fig. 127). The availability of sediment is an important factor in the process, and together with the character and growth rate of the algal film, influence the rate and periodicity of sediment accumulation and the size of particles trapped.

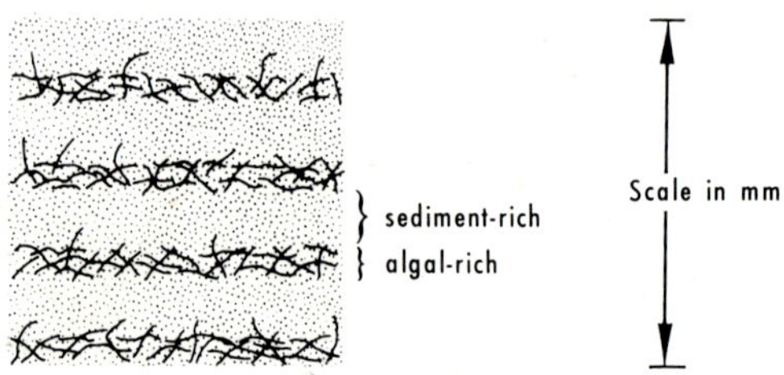

Fig. 127. Diagrammatic sketch illustrating interbedded organic films and sediment in algal-laminated structures.

Blue-green algae are cosmopolitan in their distribution. They have physiological and cytological characteristics that permit them to live in harsh intertidal and supratidal environments, as well as grow successfully in deep water (Monty, 1971). These algae are able to survive exposure during low tide because they are resistant to drought, high temperature, and solar radiation. The photosynthetic capabilities and spectral requirements of blue-green algae permit them to range to depths of over 100 m.

The more obvious biosedimentary products of nonskeletal blue-green algae are distinctly laminated. Less evident, but equally important, are sediments in which the influence of algae is more commonly inferred than observed. Aitken (1967) proposed the useful term *cryptalgal* to define sedimentary structures believed to originate through the sediment-binding and/or carbonate-precipitating activities of nonskeletal algae. Although he included stromatolites and oncolites in this category, the term seems more applicable to biosedimentary structures lacking distinct laminations, such as macroscopic clotted fabrics.

Birdseye structures (fenestral fabrics) result from several different processes, including algae, dessication, burrowing, gas genesis, and diagenesis (Deelman, 1972). The combination of algal mats and associated gas pockets seems to be the origin for many birdseye fabrics. Nonskeletal blue-green algae, which in other circumstances might form laminated structures, are no doubt a principal cause of these kinds of cryptalgal fabrics (Fig. 129).

CHARACTERISTICS OF ALGAL STROMATOLITES

Morphology and classification

The fundamental characteristic of a stromatolite is the lamination. Kalkowsky (1908) used this as a basis for defining the term and subsequent workers, for example Cloud (1942), Logan et al. (1964), and Hofmann (1969, 1973), have emphasized the importance of this feature.

Laminae result from the periodic or episodic nature of the accretion process. In Recent stromatolites, one may differentiate alternating layers of organic material and sediment. In the geologic record, this is translated into a simple succession of laminae usually defined by textural differences.

Widespread and uninterrupted algal mats cause more or less planar laminations without relief, whereas sediment accumulation on areally restricted algal films produces fixed structures of successive laminae with definite relief and boundaries, namely stromatolites (Fig. 128). Oncolites are a kind of unattached or mobile stromatolite characterized by concentric laminations (Fig. 130). The form and size of stromatolites and oncolites reflect various environmental controls.

Stromatolites often develop as discrete columns and domes; however, their gross morphologies range from low-relief mounds, through bulbous forms, to long, slender columns, and with various styles of branching (Figs. 131, 132, 133). Always, their intrinsic characteristic is internal lamination, which usually shows an inheritance of the form of earlier laminae.

Stromatolites vary greatly in size, ranging from a few millimeters to very large-scale mounds. Most often they are measured in centimeters or a few tens of centimeters. In the extreme, Hoffman (1967) reported structures 80 m long, 45 m wide, and 20 m high, with up to 2 m of relief in Proterozoic carbonates of the Great Slave Lake area. Ancient stromatolites generally have millimeter-scale laminations, although small columns with much finer laminations are not uncommon (Fig. 134).

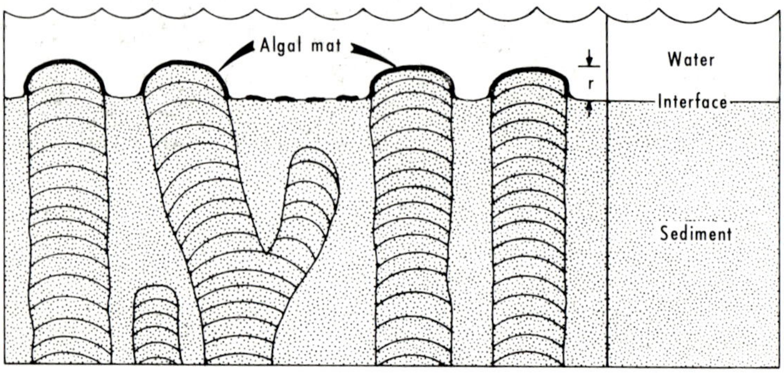

Fig. 128. Schematic diagram illustrating development of columnar stromatolites. At water-sediment interface restricted areas of algal mats localize sediment accumulation and perpetuate upward growth of columns. r = depositional relief compared to overall height of stromatolite structures. (After Hofmann, 1969).

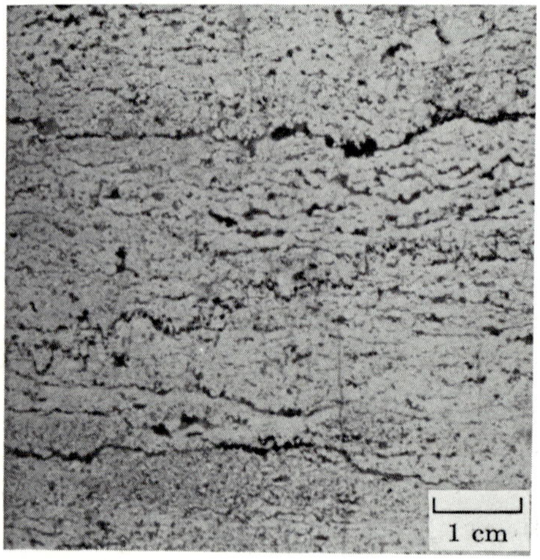

Fig. 129. Birdseye fabric. Sparry calcite (dark areas) fills original voids. Permian, Wyoming. Reflected light, polished surface.

Fig. 130. Oncolites. Whole structure (above) and section showing concentric laminations (below). Upper Carboniferous (Pennsylvanian), New Mexico.

Fig. 131. Columnar stromatolites. Cambrian, Alberta. Reflected light, polished surface.

Fig. 132. Lacustrine stromatolites; branched habit. Eocene, Wyoming. Reflected light, polished surface.

The time significance of laminations in ancient stromatolites is not clear (Park, 1976). In areas of Recent stromatolite development, Monty (1967) and Gebelein (1969) recorded algal growth and sedimentation that follow a diurnal pattern resulting in the formation of a laminae pair. On the other hand, some laminations record longer intervals of time representing seasonal, annual, or other regular cycles. It has also been observed that the production of laminations is intermittent and related to storms or other irregular happenings. Thus, the calendar nature of stromatolites is complicated because the number and thickness of stromatolite laminae may represent a wide range of time events at various regular and irregular intervals.

Early workers gave generic and specific names to different stromatolite forms and this practice, although modified, is still followed by some, especially in the U.S.S.R. *Cryptozoon* and *Collenia* are common examples of these "form genera". Cloud (1942) questioned this practice, noting that, although stromatolites are organic in origin, the structures are themselves neither organisms nor parts of organisms. It is now generally appreciated that algal mats forming stromatolites are composed of numerous species belonging to several genera, and that environmental factors have played a major role in controlling the growth form; thus the use of a biological nomenclature seems inappropriate. However, many workers dealing with

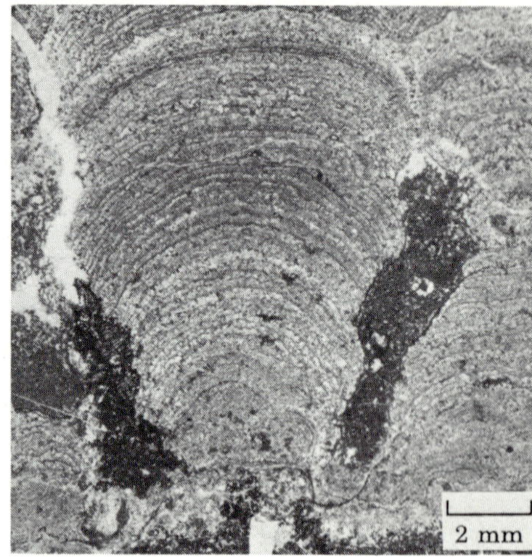

Fig. 133. Columnar stromatolites. Relief on top of limestone block approximates depositional relief on seafloor during stromatolite formation. Devonian, Western Australia.

Fig. 134. Vertical section of small columnar stromatolites about 1.5 cm high showing thin, distinct laminae. Upper Carboniferous (Pennsylvanian), Colorado. Transmitted light, thin section.

Precambrian stromatolites maintain that their distinctive morphologies can be "named" and treated as if they were in fact organisms.

Various descriptive classifications have been applied to stromatolites. For example, Donaldson (1963), using descriptive adjectives, recognized six basic forms in a study of Proterozoic structures from Newfoundland. A very different classification scheme, proposed by Logan et al. (1964), utilizes a variable descriptive formula of letter symbols based on shape, linkage, and stacking.

Hofmann (1969) presented a thorough analysis of the attributes of stromatolites, considering all aspects of their internal and external geometry, and made a comprehensive review and comparison of previous stromatolite classifications. In this review, he suggests a classification based on precisely defined geometric characteristics. In the final analysis, however, stromatolite classifications are still a controversial matter and no single approach to the matter is generally accepted.

Recent stromatolites

The present understanding of ancient stromatolites is due largely to knowledge gained from studies of Recent analogs. Had this approach not been followed, interpretations of these structures in the geologic record would be much less clear.

Black's (1933) investigation of Recent laminated algal structures in the Bahamas was the beginning of the modern era of stromatolite synthesis. He recognized the complex nature of the algal assemblages found in intertidal mats and showed that laminations were developed by the alternation of sedimentation and the growth of algal films.

Ginsburg (1955, 1960) expanded Black's work and emphasized the analogy between ancient stromatolites and Recent stromatolitic sediments in Florida and the Bahamas. Ginsburg's conclusions and influence were instrumental in fostering further studies of stromatolites at a time when investigations of other aspects of carbonate sediments were expanding.

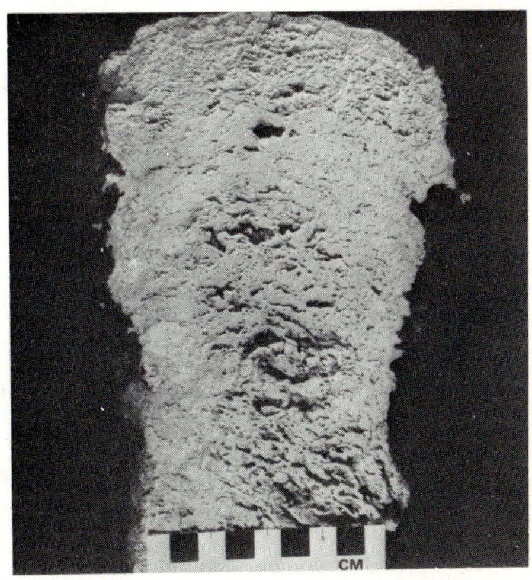

Fig. 135. Group of intertidal stromatolites. Scale in lower right 10 cm long. Shark Bay, Western Australia. (Courtesy of P. E. Playford).

Fig. 136. Vertical section of columnar stromatolite showing laminations. Shark Bay, Western Australia.

The large clublike stromatolites in Shark Bay in Western Australia have provided the only close analogy to the columnar stromatolites typical of Precambrian sediments (Logan, 1961). Developed mainly along current-swept headlands, these structures are made up of coarse laminae composed of sand-sized particles (Figs. 135, 136). The hypersalinity of Shark Bay is a major factor in the development of algal mats and stromatolites, because the condition inhibits competition from marine invertebrates. Later studies by Logan et al. (1974) have recognized different types of algal mats, reflecting variations in algal species present, and the distribution of mat types as a function of environmental factors.

Less dramatic Recent algal stromatolites, but nonetheless widespread algal-laminated sediments, occur in the Trucial Coast of the Persian Gulf (Kendall and Skipwith, 1968).

STROMATOLITE STRATIGRAPHY

Stromatolites range from the Late Archaean; the oldest ones known (about 2.6 billion years) occur in the Bulawayan Group of Rhodesia (Schopf et al., 1971). The time interval spanned by stromatolites, especially the time range of their acme during the Late Proterozoic, is indeed impressive (Fig. 137). Stromatolites are present in various Phanerozoic sediments, but are much less prominent than in the Precambrian. Their decline in the early part of the Paleozoic has been attributed to the diversification of grazing invertebrates which both eat algal mats and burrow through them to destroy sedimentary structures (Garrett, 1970).

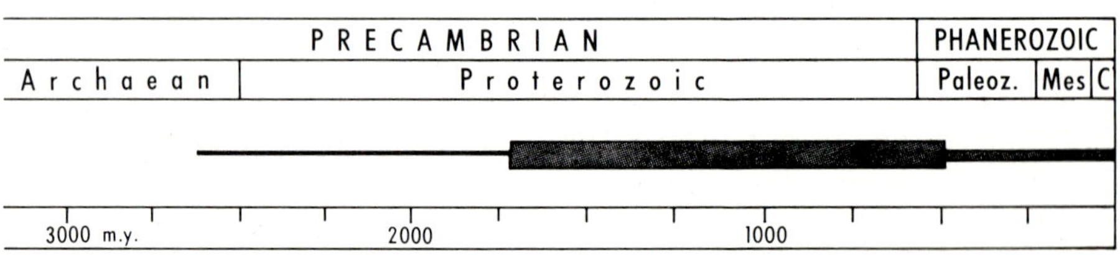

Fig. 137. Geologic range and relative abundance of algal stromatolites.

Columnar stromatolites have been used for biostratigraphic zonation in the Precambrian. Four major assemblages based on numerous form taxa have been recognized and used to subdivide the Late Proterozoic in the U.S.S.R. (see Hofmann, 1973). This chronology has been applied to other regions (Cloud and Semikhatov, 1968; Walter, 1972); however, many discrepancies in ranges and methodology have been noted, and a worldwide Proterozoic stromatolite zonation is open to criticism.

Stromatolites may be useful stratigraphic markers within sedimentary basins. Rock-stratigraphic correlations based on nonrepetitive sequences of stromatolite forms have been reported by Hoffman (1967) and Bertrand-Sarfati (1972).

ENVIRONMENTAL DISTRIBUTION

Blue-green algae have produced stromatolites most often in shallow marginal waters of marine basins and saline lakes. Recent stromatolites occur principally in supratidal, intertidal, and shallow subtidal zones. Because of these present-day associations, there has been a tendency to infer shallow-water environments for sediments containing ancient stromatolites. Playford and Cockbain (1969) have questioned this assumption on the basis of widespread columnar stromatolites in Devonian fore-reef facies in Western Australia that developed in water at least 45 m deep.

Algal stromatolites are most likely to have formed in low-latitude environments. This is not necessarily because of any ecological restriction of blue-green algal mats, but rather because marine, shallow-water, carbonate deposition predominates in these regions.

Elongated stromatolites in intertidal environments may reflect the movement of tidal currents. Recent stromatolites in Shark Bay show an elongation of elliptical, columnar structures parallel with the direction of wave scour, thus normal to the shoreline (Logan et al., 1974). Hoffman (1967) reconstructed a paleocurrent system in Proterozoic sediments based on the shape and orientation of stromatolites.

Some stromatolite forms reflect only the trend of currents, while others are inclined and asymmetric and can be used to distinguish the sense of direction, namely landward versus seaward, as well as the orientation of the shoreline.

Although most investigations of ancient and modern stromatolites suggest that elongated forms are developed normal to the shore, at least some ancient elongated stromatolites show other orientations. Young and Long (1976) described Proterozoic stromatolites believed to have formed parallel to the coastline based on paleocurrent directions interpreted from associated sedimentary structures. The elliptical shape may have resulted from tidal currents running parallel to the length of an elongated embayment in the Precambrian Sea.

Logan et al. (1974) concluded that the hydrodynamic setting of Proterozoic and hypersaline Phanerozoic shorelines can be inferred from stromatolite morphologies. Based on the Shark Bay examples, they consider discrete columnar stromatolites of relatively high relief (>20 cm), and generally circular in cross section, to be diagnostic of headlands and shores with steep gradients where wave action is most intense. Lower relief structures (<20 cm) and elliptical forms develop under conditions of moderate wave attack. Algal-laminated sediments without appreciable relief indicate environments protected from wave attack, such as shallow embayments and broad intertidal flats.

Chapter 7

CALCAREOUS ALGAE AND THEIR ENVIRONMENTS

Although calcareous algae occur in many kinds of environments, they are most abundant and varied in shallow, tropical and subtropical marine waters where the bulk of carbonate sediments accumulate today and have been deposited in the geologic past. Because calcareous algae are common constituents in many carbonate rocks and different groups and taxa of living calcareous algae have specific ecological requirements, they are useful fossils in reconstructing details of ancient sedimentary environments. Clearly, calcareous algae can contribute to an overall understanding of historical geology, and they have pragmatic value in petroleum and mineral exploration by isolating facies of prospective economic interest. The factors that limit the distribution of living algae and the observed ranges of living and fossil forms are considered in this chapter.

ENVIRONMENTAL FACTORS

The interrelationships of living algae and their environments are complex and a myriad of factors combine to limit a particular species in space. The conditions of the environment most often cited include:

- Physical factors: temperature, light, currents, substrate, physiography of the shoreline or basin.
- Chemical factors: water chemistry, dissolved salts, gases, nutrients.
- Biological factors: competition, available hosts, grazing pressure.

To better understand the nature of these controls, a few of the more important physical factors are discussed below.

Light intensity and quality

Light is a critical factor because of its role in photosynthesis. The intensity and quality of light available to marine plants at different depths depend on many variables -- reflection, scattering, absorption, transparency of the water (amount of suspended and dissolved matter), and turbulence. Depth of penetration is regulated by latitude; consequently, the more vertical are light rays, the deeper the penetration.

The depth distribution of algae is determined by the capacity of different genera and species with different pigments and metabolic processes to make use of the amount and kind of light which penetrates the water column. Algae living in low-intensity light conditions in deep water obviously have the necessary light absorbing pigments to survive. In contrast, the few algae that are subaerial part of the time, such as the blue-green algae, are protected against intense light by having darkly colored sheaths.

Clear marine waters readily absorb light rays, so that as much as 50 percent is lost in the first few meters of depth (Fig. 138). Measurements show that the amount of available light remaining at depths of several tens of meters is only a few percent.

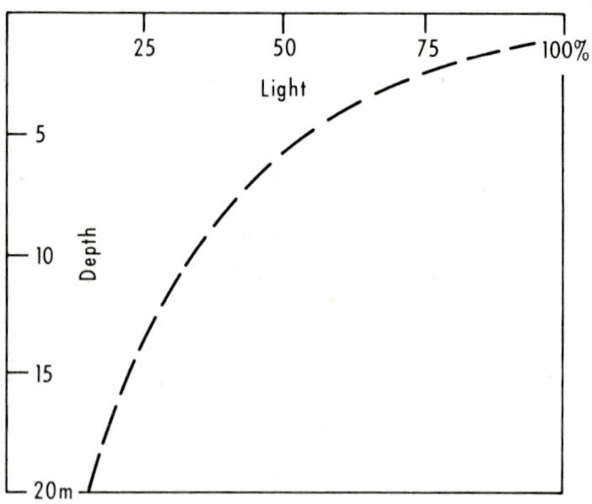

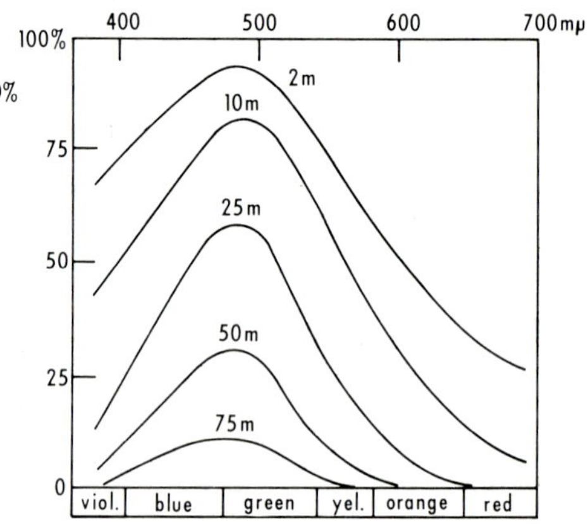

Fig. 138. Decrease in light intensity as a function of water depth. Percentage of light reaching bottom in a hypothetical coastal locality.

Fig. 139. An approximation of the spectral composition of radiant energy at different depths in clear oceanic water.

Light intensity not only diminishes with depth, but different light wave lengths are selectively absorbed in the water column, so that the longer red and orange wave lengths are absorbed first, whereas blue and green light penetrate the deepest (Fig. 139). The Chlorophyta have their maximum absorption and most efficient photosynthesis rate in the red sector of the spectrum, thus the green algae occur primarily in shallow waters. Many species of Rhodophyta have a maximum absorption in the blue sector of the spectrum, therefore the red algae are the ones that range to the greatest depths where only blue and green wave lengths penetrate.

The pigment phycoerythrin in the red algae serves as a receptor for the spectral energy available at great depths and it has been noted that the same species of red algae at great depths contains larger amounts of this red pigment than in shallow depths. There is no absolute correlation between pigments or the kind of algae and light intensity and quality, because some green algae exist in weak light at fairly great depths, and many red algae are able to adjust to intense light in the intertidal zone.

Temperature

Temperature usually is considered a less important environmental factor than light. Nonetheless, the influence of temperature is certainly evident in the latitudinal distribution of major groups of algae and in the ranges of different genera and species. For each species there is an ideal temperature range at which metabolism occurs; also, the effective use of light for photosynthesis is closely related to temperature.

Adey (1966) determined the distribution of crustose coralline species in the North Atlantic as a function of temperature and depth (light), and found that these two factors combine to define optimum living conditions for different species (Fig. 140). For many algae, temperature is a major condition in controlling growth and reproduction.

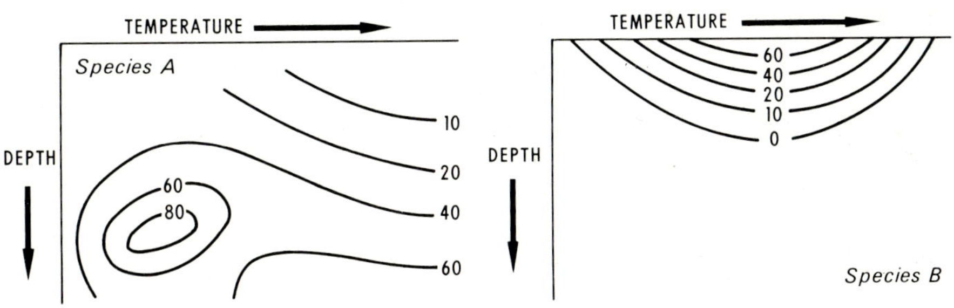

Fig. 140. Temperature and depth control of the relative abundance of two hypothetical crustose coralline species (A and B). Based on data from Adey, 1966.

Water movement

Tides, wave action, and currents effect the distribution of marine algae. Calcareous algae as a group span the entire range of the energy regime, but individually, some taxa require high-energy environments, while others survive only under conditions of gentle water movement.

The relationship of growth form and current action (and water depth) has useful implications in paleoecology (Wray, 1971). Recently, Bosence (1976), in studies on the coast of western Ireland, observed the systematic differences in form of the same species of calcareous algae in response to variations in current action (Fig. 141). With increasing exposure to bottom currents, *Lithothamnium corallioides* showed an increase in the branching density of the thallus; thus, open-branching forms are found in quiet areas and densely branched forms in exposed regions.

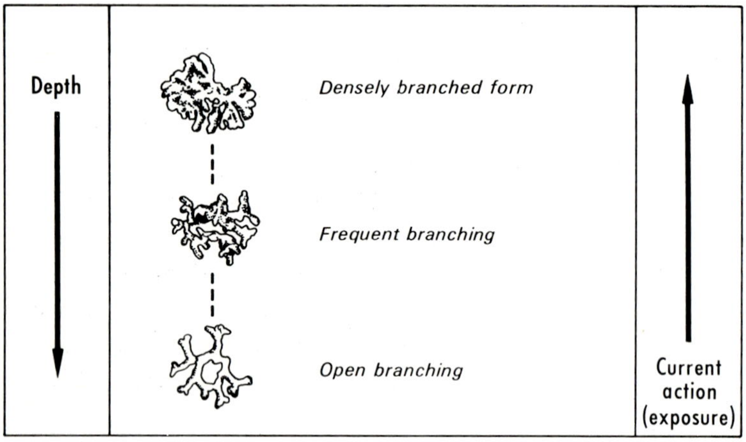

Fig. 141. Growth-form variation continuum within same species of free-living coralline as a function of depth and current action. Distribution related to depth suggested by Wray (1971); occurrence pattern controlled by current action observed by Bosence (1976).

Species of articulated corallines -- which appear to be of delicate construction, but in reality are pliable and resilient -- may prefer the pounding surf of intertidal regions, and so there is not always a correlation between growth form and intensity of water movement. The preferred energy regimes of each major group of calcareous algae was outlined in earlier chapters.

Substrate

Groups of calcareous algae show a preference for certain types of substrates, with some important exceptions. Consequently, generalized environmental associations of different calcareous algae and the character of the bottom include:

- Coralline algae: firm substrate; reefs, rocky and sandy bottoms.
- Codiacean and dasycladacean algae: soft substrates; sand and mud.
- Charophytes: soft substrate; mud.
- Blue-green algae: soft substrate; mud.

Many of the exceptions occur among the coralline red algae. Granted, all articulated forms and most crustose varieties require a firm attachment to solid objects, but an important minority of crustose corallines live free on various kinds of unstable substrates.

In studies of living North Atlantic corallines, Adey (1970b) noted that granule substrates only seldom have corallines on their surface and sandy bottoms lack these plants altogether unless larger materials are intermixed. Some crustose species show no relationship to substrate particle size, whereas *Leptophytum laeve* has a preference for finer grained particles than other crustose corallines in the environment.

Species of free-living, branched corallines may develop in such profusion as to create their own substrate of semirigid structures formed by the branches of intergrown thalli. Appreciable amounts of mud and other fine-grained sediments may be trapped within this network so that a muddy matrix is formed. Recent banks of these kinds of coralline accumulations have been described by Ginsburg and James (1974) in Florida and by Bosence (1976) in Ireland.

DISTRIBUTIONAL PATTERNS

The favored habitats of living marine calcareous algae provide an important basis for environmental interpretations of ancient sedimentary rocks containing identifiable fossil algae. Benthonic calcareous taxa develop in greatest abundance and diversity in nearshore environments, usually in water depths of a few meters to a few tens of meters (Fig. 142). The crustose corallines and the Codiaceae include a few deep-water forms which may be locally common. Filamentous and coccoid blue-green algae, which produce algal-laminated sediments and stromatolites, are most common in near

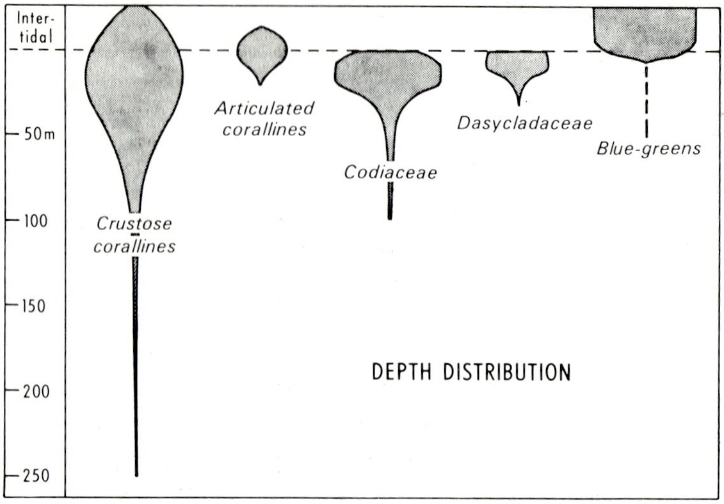

Fig. 142. Generalized depth distribution (abundance and diversity) of major groups of living marine calcareous algae.

intertidal regions, although some species have formed biosedimentary structures in deeper water marine environments.

Tropical marine waters contain the largest number and diversity of calcareous genera with all major groups represented (Fig. 143). Towards higher latitudes, the Codiaceae are the first to disappear, followed by the Dasycladaceae. Articulated corallines persist to subarctic waters, while crustose corallines are the only group of benthonic calcareous algae in very high-latitude, cold-water regions. Planktonic calcareous algae, the coccolithophorids, are cosmopolitan, ranging from the tropics to the polar seas; however, they are most abundant and diverse in warm-water, oceanic realms. Circulation and other factors may alter the latitudinal distribution of species, so that the simplified patterns outlined here may be modified.

Marine carbonate sediments are accumulating today principally on the continental shelves and shelf margins north and south of the equator below latitudes of about 30 degrees. Within this favored realm of carbonate sedimentation, benthonic calcareous algae are conspicuous biotic elements along with various invertebrate constituents. An idealized profile of a low-latitude carbonate shelf shows a preferred distribution of individual groups of calcareous algae (Fig. 144). Tidal-flat environments in both marine and nonmarine settings are characterized by algal-laminated sediments and stromatolites formed by blue-green algae, although these algae occur and may produce biosedimentary structures in subtidal realms. Charophytes

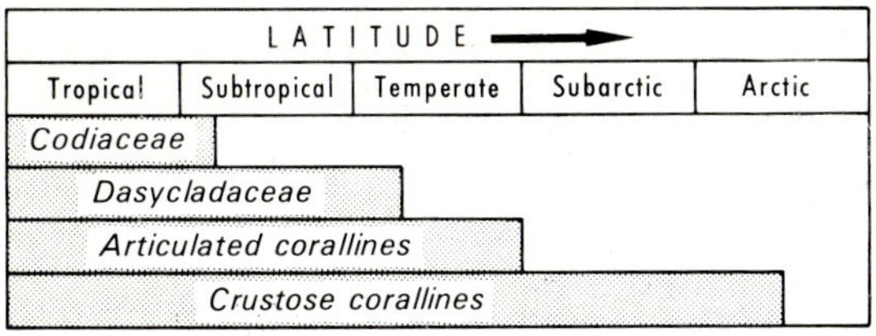

Fig. 143. Generalized latitudinal distribution (temperature regions) of major groups of skeletal benthonic calcareous algae.

accumulate primarily in lacustrine settings, but may be transported and deposited in nearshore marine environments. Calcareous codiacean and dasycladacean green algae are most abundant in relatively shallow, protected lagoons. Calcareous red algae preferentially occur in reefs, shoals, and banks, but also range into deeper waters. Coccolithophorids are indicative of open-marine conditions, but may be carried inshore and deposited in shallow-shelf sediments. The overall spatial distribution of these groups may be rather broad, but occurrence maxima tend to be restricted to particular environments.

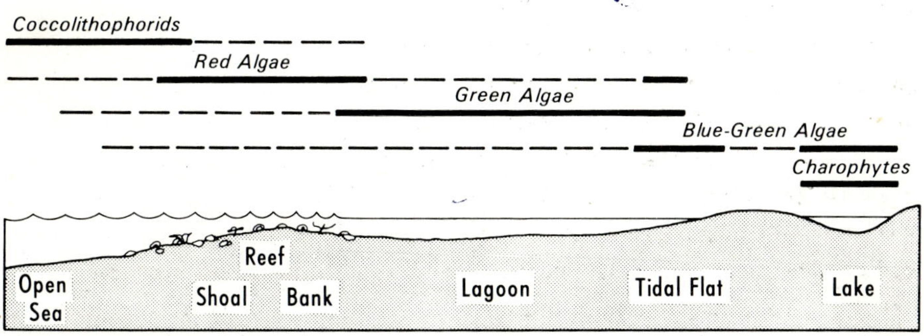

Fig. 144. Distribution of living skeletal and nonskeletal calcareous algae (including calcareous plankton) across an idealized profile of a carbonate shelf and adjacent lacustrine environments in a low-latitude region.

ECOLOGICAL SURVEYS

Much is known about the vertical distribution of the total algal flora in littoral and shallow sublittoral environments, especially in areas of appreciable tidal range. Numerous vegetation zones or belts, each characterized by a unique plant assemblage, develop in response to tide levels, wave heights, exposure, and other factors. Different kinds of plant associations occur in different climatic regions. Intertidal zonation of marine algae is often dramatic, particularly on near-vertical surfaces, and illustrates the sensitivity of these plants to varying environmental conditions. However, these kinds of surveys of living algae have produced very little information on calcareous floras that have application to the interpretation of ancient sedimentary environments where a knowledge of broader aspects of paleobathymetry and other conditions of deposition are desired.

The most significant and comprehensive studies in the field of algal ecology useful to paleoecologists have been carried out by Walter Adey and consist of numerous surveys of crustose coralline algae throughout the North Atlantic and in the Pacific. At the generic and specific level he has systematically recorded depth, temperature, and other factors of coralline floras in numerous regions ranging from the Arctic to the tropics. From this work definitive depth-distribution patterns of living taxa have been established that have value in interpreting the paleobathymetry of earlier Cenozoic crustose coralline assemblages in comparable temperature regimes.

Fig. 145 illustrates the pattern of depth distribution of major genera in the Hawaiian Archipelago (Adey and Boykins, in press). *Porolithon, Neogoniolithon,* and *Paragoniolithon* are dominant in the intertidal, reef flat, and shallow water environments (<10 m). *Hydrolithon* and *Lithophyllum* are important at intermediate depths of 10-40 m, and *Mesophyllum* and *Lithothamnium* are most significant at depths greater than 50 m. Although many of these genera have broad depth ranges, the percentage composition of the total crustose coralline flora characterizes particular depth zones.

Late Cenozoic crustose corallines have evolved slowly under tropical conditions. For this reason, they have little value for stratigraphic correlations, at least for periods of 10-20 m.y., but this enhances their value for paleoecological purposes. The present-day Hawaiian crustose coralline genera and species assemblage probably has existed unchanged since Miocene time. A crustose coralline flora of similar composition and depth distribution occurs in the Caribbean. Reefs develop under high energy conditions in this Atlantic region and they build massive algal ridges not unlike those in the Pacific (Adey, 1975; Adey and Burke, 1975).

By using parameters developed by Adey and his coworkers it is now possible to make detailed interpretations of Neogene paleoenvironments using the relative abundance of identifiable crustose genera found in outcrops and subsurface cores. At least five depositional environments -- lagoon, back-reef flat, shallow reef or algal ridge (<30 m), mid-depth

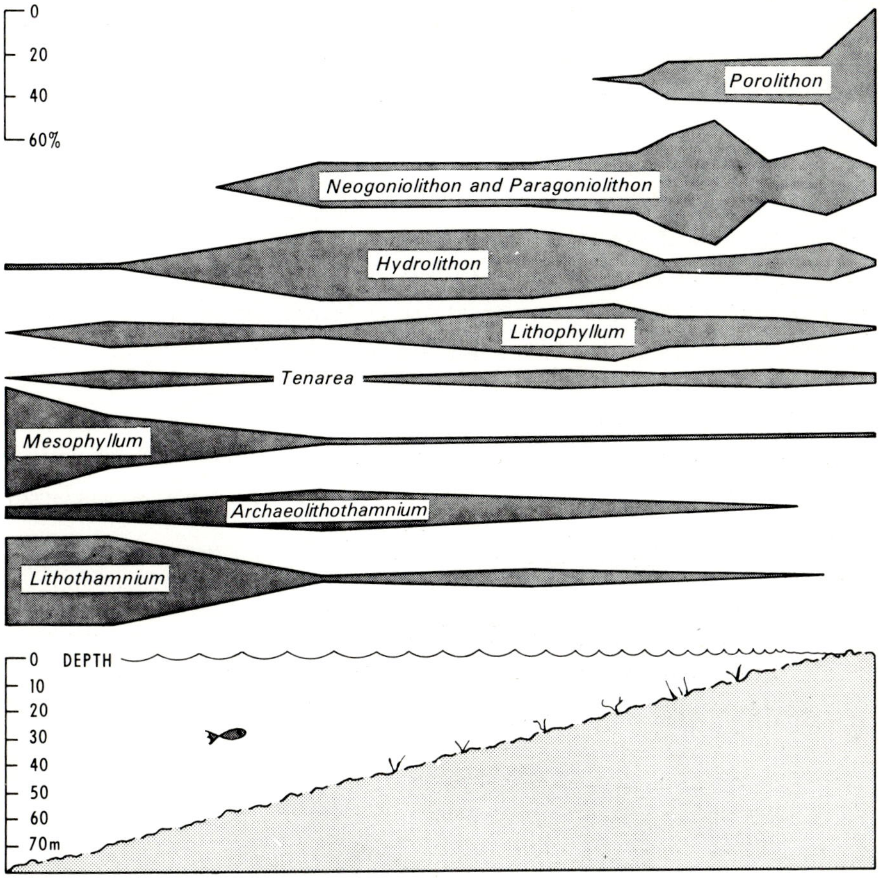

Fig. 145. Depth distribution and percentage abundance of crustose coralline genera in the Hawaiian Archipelago. (After Adey and Boykins, in press).

fore reef (30-50 m), and deep fore reef (50-150 m) -- should be differentiable. Even more precise paleoenvironmental determinations could be based on the distribution of certain distinctive coralline species with narrowly defined ecological niches; however, species identification of fossil corallines in thin section is not always possible because diagnostic features may not be preserved.

In an earlier chapter it was pointed out that the distribution of *Halimeda* species is highly depth dependent. Goreau and Goreau (1973) determined the bathymetric distribution of individual species of *Halimeda* in Jamaican reefs (Fig. 146). Moore et al. (1976) confirmed this distributional pattern and have used the occurrence of *Halimeda* species in sediments from fore-reef slopes as a means of determining the dispersal patterns of shallow-reef sediments into deeper depositional environments. From this they conclude that *Halimeda* is not only a major reef-sediment source but also a useful sediment tracer.

Utilizing *Halimeda* species for paleoecological interpretation of carbonate rocks poses problems. Although some living species may be easily recognized on the basis of the shape and size of segments, this may not be possible in thin sections of fossils in lithified sediment. In addition, the ease with which *Halimeda* segments are transported downslope masks their real occurrence patterns and only the deepest water indicators could be used.

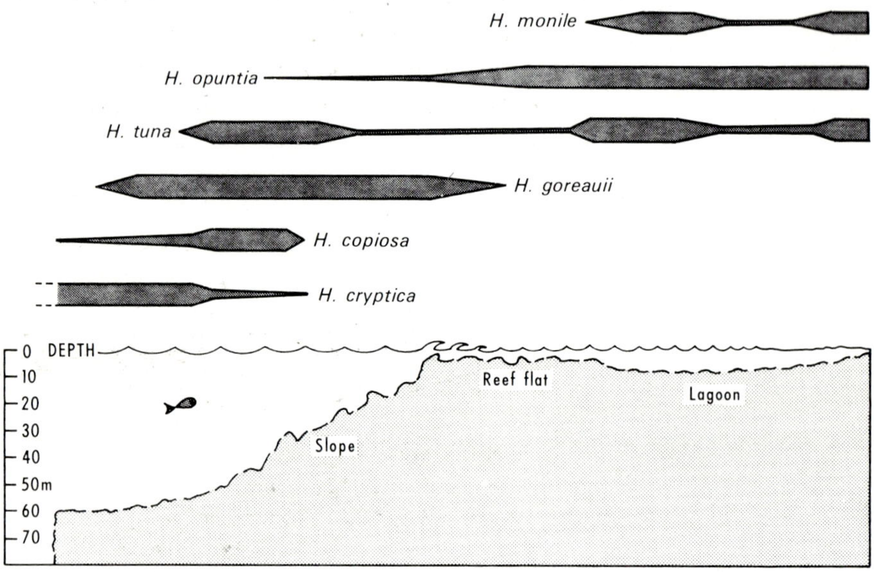

Fig. 146. Depth distribution and relative abundance of some *Halimeda* species in Jamaican reefs. (After Goreau and Goreau, 1973).

ANCIENT SEDIMENTARY ENVIRONMENTS

Interpreting Paleozoic depositional environments using calcareous algae depends more on establishing empirical distribution patterns than on utilizing ecological zonation of living algae because many ancient taxa are extinct and their relationship to living forms may be unknown. Nevertheless, Paleozoic calcareous algae can provide indices of paleoenvironments that complement interpretations based on other biota and sedimentary features. The following two examples illustrate the range of possibilities.

Devonian reef complexes

Calcareous algae are widely distributed in most of the major facies of Upper Devonian carbonate reef and bank complexes of Australia, Canada, and the U.S.S.R. (Wray, 1972). Many taxa are quantitatively significant sediment contributors, and some encrusting forms had a role in reef-building by erecting self-supporting skeletal frameworks and in binding sedimentary particles. Although many genera have uncertain biological affinities, they are nonetheless useful in paleoenvironmental analyses.

The observed facies distribution of the principal kinds of skeletal calcareous algae in Upper Devonian reef and bank complexes is illustrated schematically in Fig. 147.

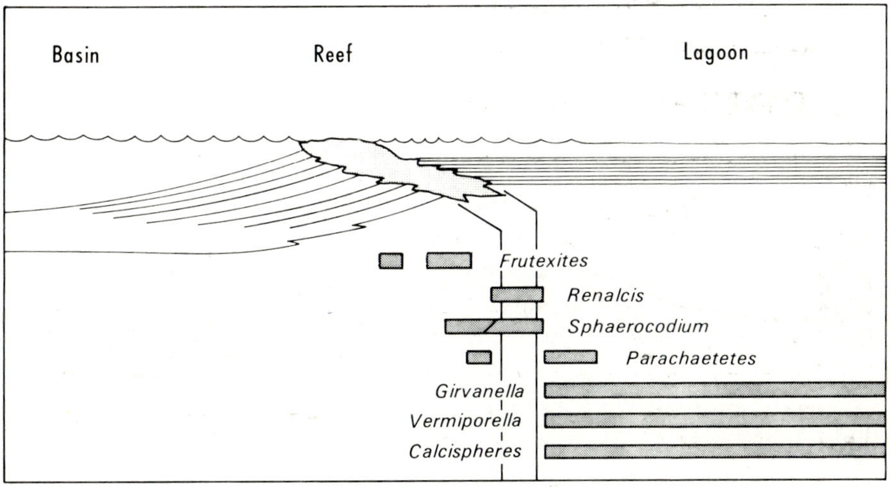

Fig. 147. Environmental distribution of skeletal calcareous algae in Upper Devonian reef and bank complexes of Western Australia and Alberta.

Shelf facies, which formed in lagoons or behind reefs, are characterized by three principal elements, radiosphaerid calcispheres, *Vermiporella* segments, and *Girvanella*. *Girvanella* occurs as nodules and laminated crusts on other constituents. Calcispheres, presumably reproductive bodies of dasycladacean algae, are widespread in this environment. *Parachaetetes* is found in some localities in facies immediately behind the reef.

Renalcis and *Sphaerocodium* are abundant in reef or bank-edge facies and these two genera are principal framebuilders in the Australian reef complexes. *Renalcis* seems to be an unusually good indicator of reef development because of its common occurrence and close association to these facies.

The indigenous calcareous algal flora in fore-reef environments consists of occasional Solenoporaceae (*Parachaetetes*) and deep-water species of *Sphaerocodium*. Fore-reef facies may contain storm-transported algal constituents indigenous to reef and lagoonal environments, in addition to allochthonous debris deposits of lithified algal limestone derived from reef and bank margins. *Frutexites* is locally abundant in deep-water stromatolites and other algal-laminated sediments which formed further down the slope and in the basin.

Remains of calcareous algae are sufficiently diverse, abundant, and widespread in Upper Devonian carbonate complexes to provide a proximity and predictive tool for outlining the distribution of depositional facies. Major elements of the algal flora are cosmopolitan, occurring in Australia, Canada and Soviet Union, although a few forms may be provincial and unique to certain regions.

Upper Carboniferous (Pennsylvanian) algal banks

The late Paleozoic red algae *Archaeolithophyllum* and *Cuneiphycus* are potentially important fossils for interpreting sedimentary environments, because they occur in abundance and appear to be restricted to specific depositional facies. No doubt individual taxa of Paleozoic benthonic red algae had very definite ecological tolerances, but much empirical distributional data are required to develop these patterns.

Paleozoic calcareous red algae seems to present a difficult problem in paleoenvironmental studies because high-order taxa of living red algae have broad ecological ranges; only genera and species show ecological restrictions. Consequently, valid generic and specific concepts of Late Paleozoic taxa must be developed to have value in paleoenvironmental interpretations.

The majority of Late Paleozoic calcareous red algae grew most commonly in open marine, carbonate-shelf environments. As a group, these algae tolerated a variety of current velocities ranging from quiet water, in which dominantly mud-rich facies were deposited, to agitated environments characterized by grain-supported fabrics with little mud-sized carbonate and in which skeletal grains are generally fragmented and abraded.

Heckel (1975) noted that Pennsylvanian Solenoporaceae occur preferentially in mud-free grainstones deposited in agitated environments. On the other hand, delicately branching forms, such as *Ungdarella* and *Cuneiphycus*, have been observed most often in mud-rich rocks suggesting quiet water deposition.

Archaeolithophyllum occurs in both mud-rich and mud-free facies, but was predominantly an encrusting alga and served as a framebuilder in the development of reefs and carbonate banks (Wray, 1964). These kinds of algal accumulations may be analogous to the Recent banks composed of the coralline alga *Lithothamnium corallioides* described by Bosence (1976).

Kotila (1973) suggests that two species of *Archaeolithophyllum*, *A. missouriense* and *A. lamellosum*, occupied different ecological niches on Pennsylvanian algal banks (Fig. 148). *A. lamellosum* occurs on the more turbulent-water crests of the bank, whereas *A. missouriense* appears to be restricted to less turbulent environments. These differences, plus the association patterns of other algae, including *Cuneiphycus*, dasycladaceans, and *Girvanella*, illustrate the potential value of these plant fossils in paleoenvironmental reconstructions of ancient rocks.

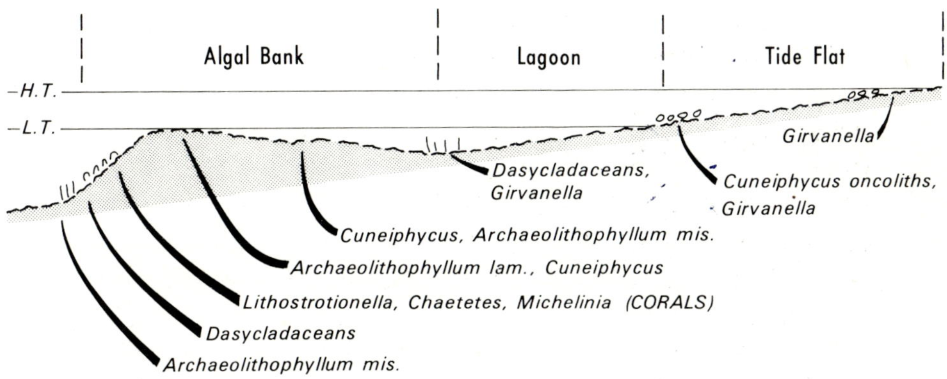

Fig. 148. Observed distribution of calcareous algae and corals in algal bank complex. Late Carboniferous (Pennsylvanian), Oklahoma. (After Kotila, 1973).

Although the variety of facies occurrences of Late Paleozoic calcareous red algae and other types suggests that individual taxa can provide important paleoenvironmental clues, quantitative distributional data must be obtained from many carefully studied stratigraphic sections within a given region.

This information should be interrelated with knowledge of other kinds of fossils, in addition to sedimentary structures and fabrics. The resulting observed associations of particular kinds of calcareous algae and specific facies can decipher paleoecological factors, such as water depth, current energy, and turbidity. This empirical approach is believed to be more promising than one relying on comparative studies of living skeletal algae and presumed ancient counterparts because of their uncertain biological relationships.

Lower Cenozoic carbonate platforms

The most important attribute of Cenozoic skeletal calcareous algae is that the assemblages are modern; that is, with only a few exceptions they have living descendants. Consequently, both identifications and environmental determinations are rather exact because we can refer to the Recent record. This is in contrast to Paleozoic calcareous algae where a high proportion of the assemblages lack modern descendants or even morphological counterparts.

Early Cenozoic skeletal benthonic algae are composed of representatives of three major groups (Codiaceae, Dasycladaceae, and Corallinaceae) and three minor groups (Solenoporaceae, Gymnocodiaceae, and Squamariaceae). These skeletal benthonic algae, plus products of nonskeletal algae (algal-laminated sediments and stromatolites) and planktonic calcareous algae (coccolithophorids) provide a large number of groups with which to analyze depositional environments on Cenozoic carbonate shelf margins.

The distributional patterns of these groups of calcareous algae, extending from tidal flat to basinal environments (Fig. 149), have been observed in Libyan Paleocene carbonate platform facies associated with reefs and banks (Wray, 1969).

Tidal-flat environments, including supratidal deposits, are characterized by various algal-laminated structures and stromatolites. Lagoonal and back-reef facies are dominated by codiacean and dasycladacean green algae, often to the exclusion of other kinds of skeletal constituents, and these algae are volumetrically important sand-sized sediment producers. The Dasycladaceae are presented by about eight genera. Codiaceae, namely *Halimeda* and *Ovulites*, appear to be more abundant than Dasycladaceae in more nearshore environments. The gymnocodiacean *Permocalculus* is a minor component of some lagoonal facies.

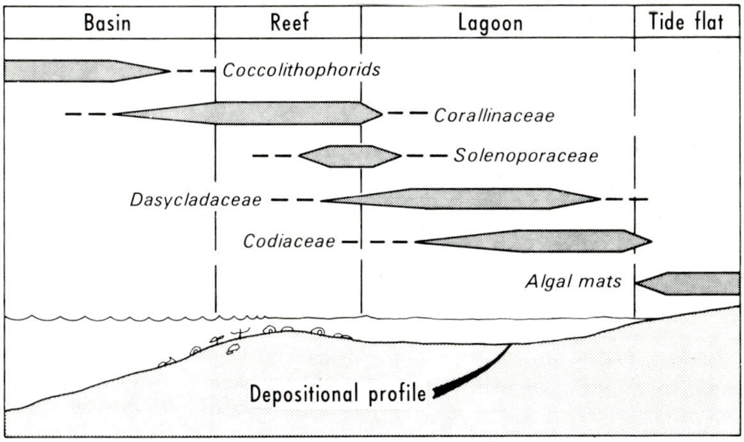

Fig. 149. Distribution of major groups of calcareous algae observed in Libyan Paleocene carbonate shelf margin.

Reefs and adjacent facies are characterized by corals and diverse kinds of skeletal algae. Several genera of crustose corallines, including *Archaeolithothamnium*, *Lithophyllum*, *Lithothamnium*, and *Mesophyllum*, are the dominant reef algae. The corallines provided an essential cementing and framework function in this environment. The squamariacean genus *Ethelia* is a minor element in reef facies. Solenoporaceans, both *Parachaetetes* and *Solenomeris*, are locally common in the transitional environment between the lagoon and reef. Corallines decrease in abundance downslope towards the basin where planktonic foraminifera and coccolithophorids are the dominant constituents.

Aspects of this generalized distributional pattern of major groups of Early Cenozoic calcareous algae have been observed in other regions and in the Middle East (Elliott, 1960, 1968). Knowledge of the facies patterns of these calcareous algae, especially when integrated with distributional data on foraminifera and other invertebrate constituents, have been keys to delineating proximity gradients to potential petroleum reservoir facies in some Cenozoic marine carbonate rocks.

Eocene lacustrine stromatolites

Because stromatolites occur commonly -- but not exclusively -- in near intertidal zones, their recognition in stratigraphic sequences, especially in association with shallow-water sedimentary structures, is a good index of shoreline environments.

Stromatolitic algal zones in the Green River Formation of Wyoming have proven useful in basin analysis of these nonmarine deposits (Surdam and Wray, 1976). Consisting of oil shale, mudstone, limestone, dolomitic marlstone, and trona, these sediments were the result of lacustrine deposition in a closed basin which was subject to repeated fluctuations in lake level and shoreline position.

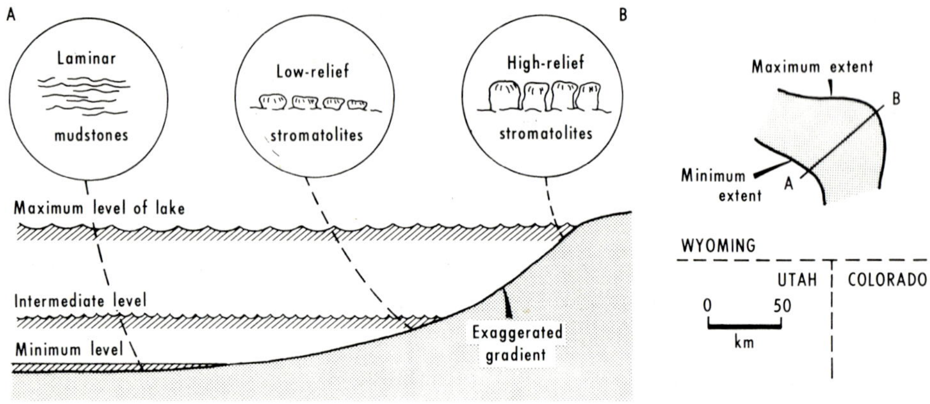

Fig. 150. Association of stromatolite form and shoreline position of Eocene Lake Gosiute. Green River Formation, Wyoming. (After Surdam and Wray, 1976).

By tracing a single stromatolitic unit from northeast to southwest (Fig. 150) over a distance of about 60 km, the stromatolites are seen to change from large individual heads up to 40 cm across (near the maximum extent of the ancient lake) to progressively lower relief forms where finally in center of the basin this same unit is characterized by mud cracks, flat-pebble conglomerates, and ripple marks. The systematic variation in the form of stromatolites within this unit is interpreted as representing a transition from a wave-swept shoreline along the northern margin of the basin at a high stand of the lake (maximum aerial extent of lacustrine sediments) to a mudflat environment at a relatively low level of the lake (a minimum aerial extent of lacustrine deposits).

Delineating the regional distribution and types of lateral changes of individual stromatolitic units is a means of determining the shoreline positions of this ancient lake basin, which has contributed to a better understanding of the sedimentology, stratigraphy, and mineral deposits of the Green River Formation.

Chapter 8

SEDIMENT-PRODUCING ALGAE

Calcareous algae as "rock-builders" were noted by the paleobotanist Seward (1898), who early recognized the volumetric significance of skeletal carbonates produced by marine algae. Although many others have since acknowledged the sedimentary contributions of algae, Johnson (1961) was very much responsible for stimulating current interest in "limestone-building algae".

In this chapter we consider the role of skeletal algae as sedimentary particles in carbonate deposition, namely:

- Sediment types produced by skeletal calcareous algae -- their origin, size, and composition.
- Volume of material.
- Distribution and dispersal.
- Extrapolation into the geologic record.

Studies of Recent carbonate shelf regions suggest that skeletal calcareous algae may be an all-important source of fine-grained sediment -- far greater than invertebrate groups and inorganic sources. Can we conclude that much of the volume of ancient limestones also owes its origin to algae? Quite possibly. Let's examine the evidence.

MODERN CARBONATE DEPOSITION

Carbonate sediments accumulate today mainly in shallow, tropical seas where both the abundance and diversity of most organisms is greatest (Fig. 151). Marine carbonates are largely dependent on some type of organic activity for the production and modification of carbonate particles. Even so-called "inorganic carbonate" sediments may be the result of biochemical or poorly understood organic activity. Because carbonate sediments are organically produced, they characteristically have a local, intrabasinal origin and are not derived from outside the basin of deposition. The regions of warm, well-lighted waters where most carbonate sediments are generated also provide optimum conditions for the growth and accumulation of skeletal calcareous algae.

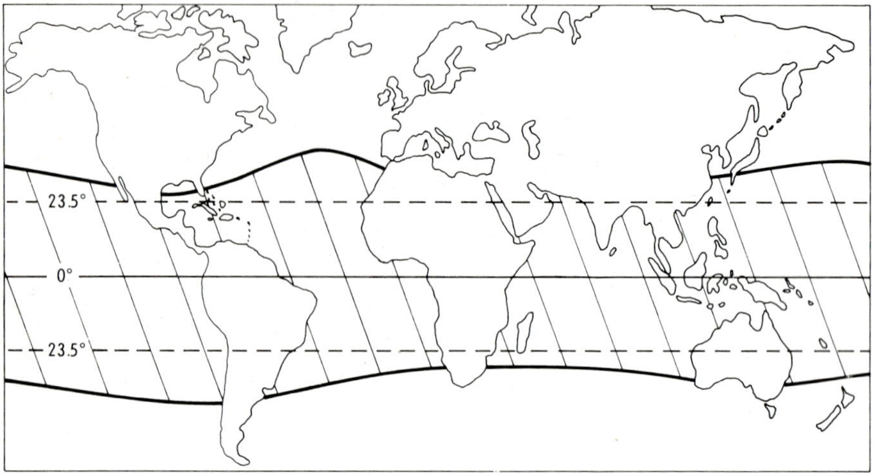

Fig. 151. Latitudinal limits of major areas of present-day, shallow-marine carbonate deposition, including reef developments.

LIVING SEDIMENT—PRODUCING ALGAE AND FOSSIL COUNTERPARTS

Coralline frameworks -- reefs and banks

The importance of modern crustose coralline algae in reef formation has long been recognized. These plants, composed of magnesium calcites, provide an essential binding and cementing function. The mass or volume of sediment these algae contribute to reefs is variable; generally, it is less than corals and the combined total of other skeletal constituents, but nonetheless a significant amount.

Values of living coralline abundance are often expressed in terms of mean cover of the surface area, but the highly irregular surface of most reefs must be taken into account. According to Dahl (1973), each square meter of the Belize (British Honduras) barrier reef crest includes about 15 m^2 of surface. Of this figure, the amount of available surface for benthonic algae, mainly corallines, is about 4 m^2. Combining data from various sources on estimates on living coralline cover, in addition to observations of Cenozoic limestones, we conclude that coralline frameworks may constitute 20-50 percent of the mass of modern reef accumulations (Fig. 152).

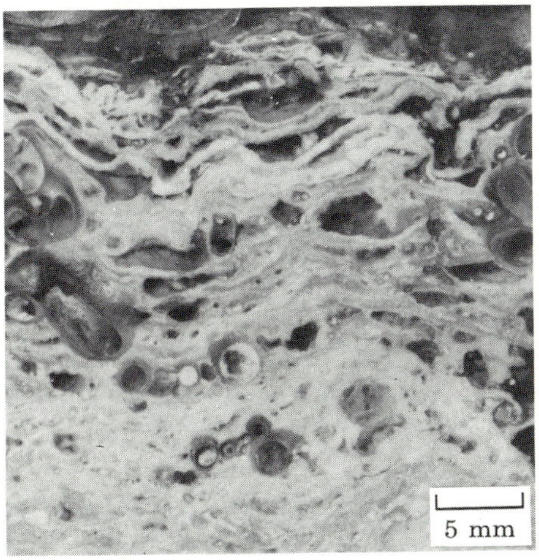

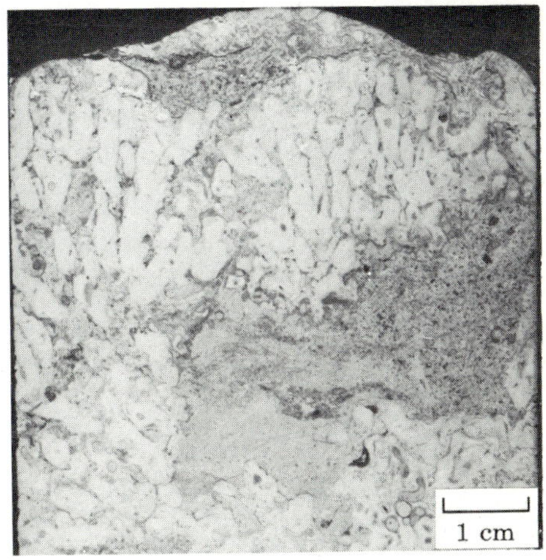

Fig. 152. Crustose coralline framework. Recent, Bermuda. Reflected light, polished surface.

Fig. 153. Framework of rigid-branched coralline algae. Paleocene, Libya. Reflected light, polished surface.

Free-living or loosely attached crustose corallines, for example species of *Neogoniolithon*, are quantitatively important producers of carbonate sediment in shallow-shelf environments, where they may construct in situ banks. These algae are often the only skeletal constituent, but because of their branched habits they are loosely packed and make up less than 25 percent of the sediment volume. Coralline frameworks of reefs and banks are prevalent throughout the Cenozoic (Fig. 153), and many fossil accumulations have attributes of Recent ones.

Paleozoic algal facies comparable to modern coralline frameworks occur in the Late Carboniferous (Pennsylvanian). The ancestral coralline alga *Archaeolithophyllum* (Wray, 1964), which occurs both as in situ crusts and as transported thalli, is the dominant constituent of these widespread and locally thick carbonate buildups (Fig. 154). In addition to its volumetric significance, sedimentological evidence suggests that *Archaeolithophyllum* provided a self-supporting, skeletal framework and sediment-binding (cementing) function which localized carbonate sedimentation.

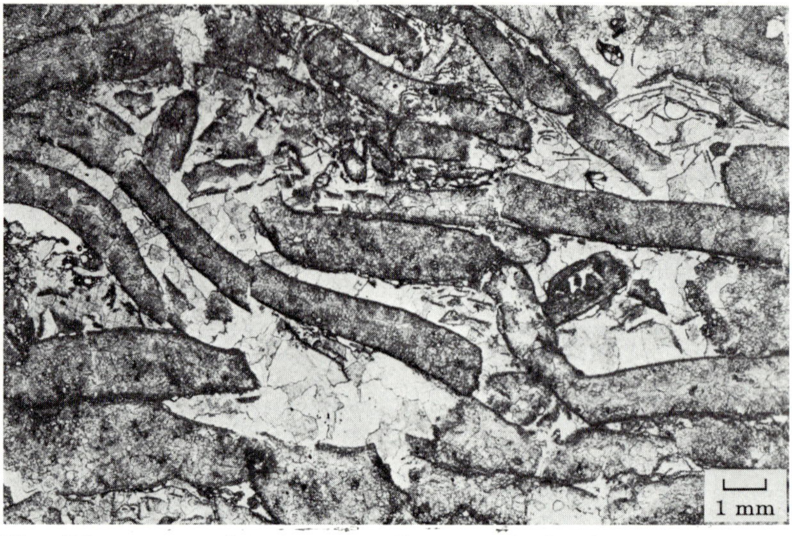

Fig. 154. *Archaeolithophyllum* grainstone of closely-packed thalli. Upper Carboniferous (Pennsylvanian), Kansas. Transmitted light, thin section.

Halimeda sands

The codiacean alga *Halimeda* is a major source of calcareous sand and larger particles. It is the single most abundant grain constituent in many subtropical lagoon and back-reef environments, often constituting 50 percent of the volume of bottom sediments (Ginsburg, 1956).

Halimeda sands also accumulate in substantial amounts in deep-water environments due to downslope dispersal of shallow-water species and locally abundant growth of this alga on the deep fore-reef slopes. Observations from submersibles (Wray, 1972a; Ginsburg and James, 1973) have documented the occurrence of substantial amounts of *Halimeda* sand at depths of nearly 300 m (Fig. 155). Similar observations by Moore et al. (1976) found that *Halimeda* grains average 25 percent of the sediment volume on some Jamaican deep fore-reef slopes.

Halimeda grainstones and packstones are common carbonate facies (Fig. 156) throughout the Cenozoic. Paleozoic counterparts to modern *Halimeda* facies are the phylloid algal limestones of Late Carboniferous (Pennsylvanian) and Early Permian ages (Wray, 1968). Typically, these facies consist of coarse-grained, platelike remains of the algae *Eugonophyllum* and *Ivanovia* in a matrix of carbonate mud and clear calcite cement (Fig. 157), although some rocks are essentially devoid of matrix material. Accumulations usually contain enough large particles of algae in contact with each other to have

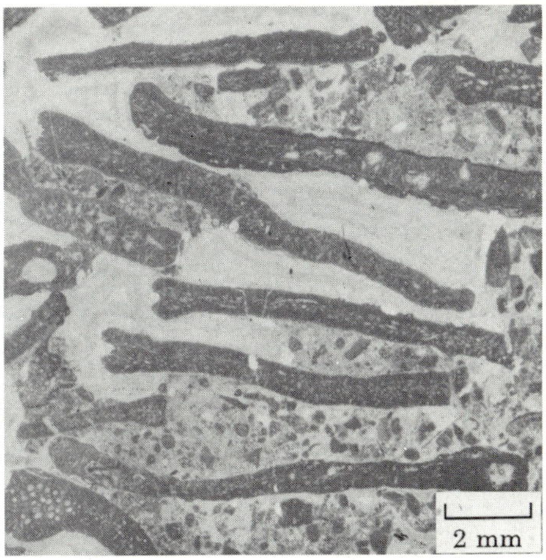

Fig. 155. Coarse-grained *Halimeda* sediment on deep fore-reef escarpment at -290 m. Individual segments approximately 1 cm in size. Glovers Reef, Belize. (Photo by J. L. Wray).

Fig. 156. *Halimeda* packstone. Pliocene, Mariana Islands. Transmitted light, thin section.

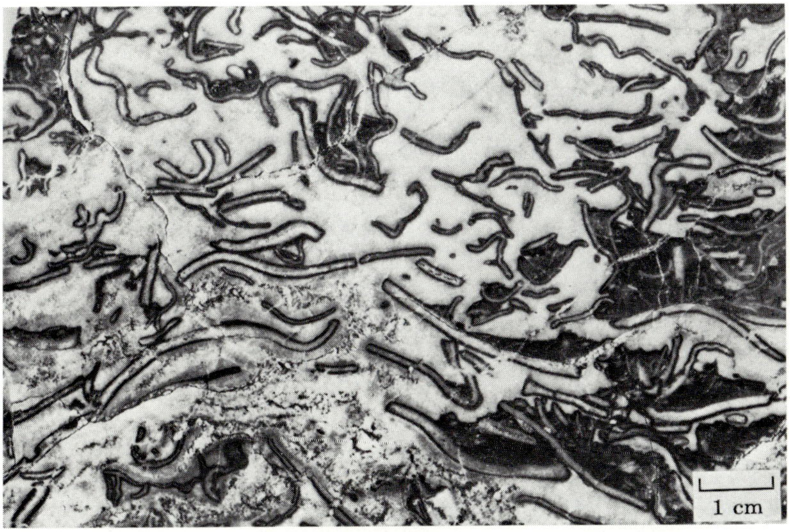

Fig. 157. Phylloid algal limestone (grainstone) consisting of micrite-rimmed algal fragments (*Eugonophyllum* or *Ivanovia*), calcite cement (white areas), and carbonate mud in sediment traps. Upper Carboniferous (Pennsylvanian), Canadian Arctic Archipelago. Reflected light, polished surface. (Courtesy of G. R. Davies).

developed grain-supported fabrics, but because of their irregular shape, they may form a framework with only 10-20 percent grain bulk.

Codiacean lime muds

Prior to the mid-1950s lime muds, both ancient and modern, were considered to be due largely to physicochemical precipitation rather than skeletal origin. Earlier Wood (1941) had suggested that some fine-grained Carboniferous limestones were the result of "algal-dust", but there was little appreciation of this process.

In 1955 Lowenstam reported that some common calcareous Codiaceae disaggregate into small aragonite needles 2-10 microns long, resembling crystals in Bahamian lime muds. To support this algal source for modern fine-grained carbonate sediments, Lowenstam and Epstein (1957) showed that the oxygen isotope ratios in the skeletons of calcareous green algae and in the sediment particles were essentially the same.

The first quantitative estimate of lime-mud production by the fragile codiacean alga *Penicillus* (Fig. 158) was made by Stockman et al. (1967). The annual contribution of fine-grained carbonate sediments from species of *Penicillus* in south Florida was determined by observations of bottom stations over a one-year period, counts of plant abundance, and measurements of the weight of aragonite manufactured by each plant. For stations in the inner reef track, annual production rates of 25 g/m^2 were calculated (Fig. 159).

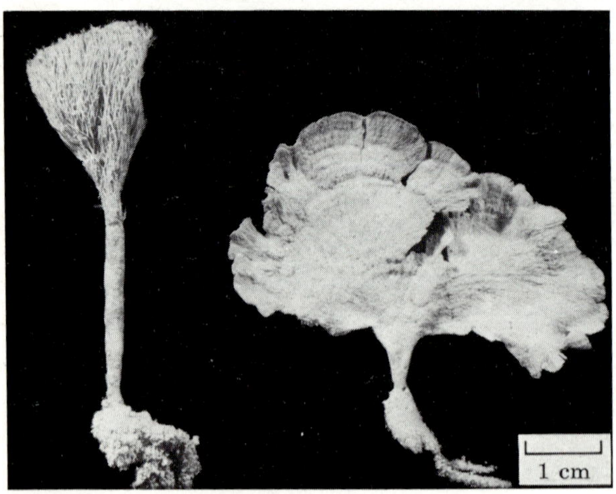

Fig. 158. Common types of erect calcareous codiacean green algae, *Penicillus* (left) and *Udotea* (right), which disaggregate into fine-grained aragonite needles.

Rates of production of about 3 g/m^2 were obtained at stations in Florida Bay. The results of this study proved that one genus alone, *Penicillus*, is a major contributor of fine-grained aragonite in amounts believed to be more than adequate to explain the accumulation of mud-sized sediments in south Florida.

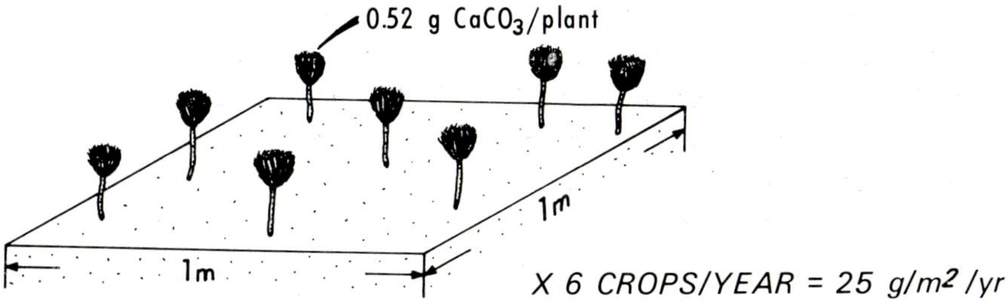

Fig. 159. Diagrammatic presentation of lime-mud production by calcareous green alga *Penicillus*. (Data from Stockman et al., 1967).

Similar studies of codiacean lime-mud production were made in the Bight of Abaco in the Bahamas by Neumann and Land (1975). In addition to *Penicillus*, they determined sediment derivation from species of related genera, *Rhipocephalus*, *Udotea*, and *Halimeda*, and estimated the magnitude of the total contribution from these algae in comparison to the amount of sediment in the lagoon basin. They calculated 90 g/m^2 per year of algal aragonite production on the lagoon floor based on a calcium carbonate contribution of 10 g/m^2 per crop and a growth rate figure of nine crops per year, which is representative of this Bahamian locality. In effect, the amount of aragonite muds produced by codiacean green algae alone is much more -- perhaps as much as three times -- the present rate of accumulation of lime-mud sediments in the lagoon basin. This overproduction of aragonite mud and *Halimeda* sand in this interior Bahamian bank lagoon becomes a source of sediment for deposition in adjacent regions, such as lateral accretion of the platform, "pelagic" deposition in the surrounding deep-water basins, and supratidal accumulations.

Substantial amounts of lime mud are produced by other kinds of algae. The encrusting coralline alga *Melobesia*, which is epiphytic on marine grasses, is estimated to yield 40 g/m^2 per year of fine-grained, magnesium calcium carbonate (Land, 1970). The brown alga *Padina* and the red alga *Galaxaura*, although lightly calcified or encased in a thin film of aragonite, may be noteworthy sediment sources in some environments.

In addition to the contribution of muds from the spontaneous disaggregation upon death and decay of the fragile calcareous algae, a considerable amount of fine-grained sediment is due to mechanical breakdown of skeletal algae.

While calcareous green algae are undoubtedly the most important single source of small elongate carbonate crystals, Stieglitz (1973) pointed out that foraminifers, mollusks, corals, and tunicates also construct similar skeletal products and that in some areas these nonalgal sources of fine-grained calcium carbonate may be significant.

Coccolith chalks -- pelagic carbonates

Cretaceous and Cenozoic accumulations of planktonic calcareous algae, the coccolithophorids, constitute quantitatively impressive masses of fine-grained, calcium carbonate sediments in relatively deep marine basins. These minute (2-25 microns) skeletal algae resist solution because of their low-magnesium calcite composition. High productivity rates and general lack of other calcareous constituents, except planktonic foraminifera, accounts for the relatively pure accumulations of coccolith sediments.

Coccolithophorid production alone provides several grams of calcite to a square meter of sea floor per year, some of which is in the form of overgrowths or reprecipitated calcite on coccoliths. Honjo (1976) estimates the annual coccolith production in the equatorial Pacific to be approximately 8 g/m^2 and it is at least the same order of magnitude as the carbonate sedimentation rate in the Atlantic.

Fig. 160. Upper Cretaceous (Maastrichtian) chalk, The Netherlands. Scanning electron micrograph.

THE ANCIENT RECORD

The sediment contributions of some skeletal algae in the geologic record are easily recognized because they have left identifiable remains. Consequently, the amount of material can be measured, and they also provide clues to the depositional environment and age of the rock. But volumetrically, the bulk of ancient carbonate sediments consists of very fine-grained particles, either in thick uninterrupted masses or as matrix among coarser-grained particles, and the origin of this carbonate mud cannot be determined directly.

The preponderance of evidence from recent studies suggests that codiacean algae are the principal source of lime muds in present-day, warm, shallow-marine environments. But can we extrapolate these findings to the origin of ancient carbonate muds in Mesozoic and Paleozoic times? *Boueina*, a genus similar to *Halimeda*, first appeared in the Jurassic. Thus, we could logically conclude that biologically related, bushy-headed, easily disaggregated codiaceans, like *Penicillus*, may have coexisted with *Boueina* and related genera in the Mesozoic and Cenozoic but did leave identifiable skeletal remains.

For the Paleozoic we can only speculate as to the role codiacean algae had in lime-mud production. Because skeletal codiaceans, such as *Palaeoporella* and *Dimorphosiphon*, lived in early Paleozoic time, there is reason to believe that fragile calcareous codiaceans may also have been present, but like *Penicillus*, they disaggregated completely into mud-sized, carbonate particles. Perhaps their remains accumulated at rates comparable to lime-mud production in modern shelf environments, and their record is the huge volume of fine-grained Paleozoic carbonate rocks.

Chapter 9

ALGAL FACIES IN TIME

Although many groups of algae have evolved slowly, very different assemblages of skeletal algae are found in rocks of different ages. The purpose here is to examine the kinds of algae -- major groups and important genera -- occurring in carbonate facies throughout the Phanerozoic. This is not to review the biostratigraphy of calcareous algae, but rather to place them in a broad time framework to compare similarities and note differences.

Often the concern of students and practicing geologists is focused on strata of a specific age. Consequently, knowledge of the groups of algae likely to be encountered in particular carbonate facies, and their significance, is a useful starting point. This resume concentrates on marine skeletal calcareous algae and only briefly touches on the chronologic distribution of stromatolites. Illustrated reviews of the calcareous algae found in many of the geologic periods were compiled by Johnson (1961a, 1963, 1964b, 1966, 1969) and Johnson and Konishi (1956, 1958, 1959).

CHARACTERISTIC AGE ASSEMBLAGES

Cambrian

The earliest known skeletal algae, cyanophytes and Dasycladaceae, appeared in Early Cambrian time; these were followed by the Solenoporaceae in the Middle Cambrian and the Codiaceae in the Late Cambrian. Thus, by the end of the period, several major groups were represented.

The dominant skeletal algae in Cambrian carbonate buildups are blue-green algae, especially *Girvanella, Renalcis,* and *Epiphyton.* Species of *Renalcis* and *Epiphyton* are not only volumetrically significant constituents, but also provided a skeletal framework. These two genera occupied similar ecological niches in the Late Devonian where they also developed in large numbers; however, they are essentially unknown from the Ordovician and Silurian, except for occurrences in the lowermost Ordovician.

Stromatolites are relatively common and widespread in the Cambrian. The well known Upper Cambrian accumulations in New York consist of large heads up to 1 m across and occur in a trend regarded as a barrier reef (Goldring, 1938). Upper Cambrian buildups in central Texas, long considered to be composed mainly of nonskeletal stromatolitic algae, were constructed primarily by *Girvanella* and *Renalcis* (Ahr, 1971).

Ordovician

During the Ordovician diverse assemblages of dasycladaceans and codiaceans evolved. Some of these taxa are unequivocal members of these families of green algae because they have definitive morphological features of modern taxa. Others, however, such as the receptaculitids and *Nuia*, have questionable biological relationships to these groups. Skeletal red algae continue to be represented by the solenoporaceans. The simple filamentous blue-green alga *Girvanella* was joined in the Ordovician by structurally complex, branched, filamentous genera.

Fragments of the problematical codiacean *Nuia* are important constituents in Lower Ordovician mound-like structures. Although both *Renalcis* and *Nuia* coexisted in earliest Ordovician time, they have not been found together in the same facies, according to Toomey and Klement (1966). *Renalcis* characteristically occurs in organic frameworks, while *Nuia* is found in clastic bank accumulations. This relationship most probably reflects different ecological requirements.

Middle Ordovician mounds and reefs in eastern United States and Canada contain *Solenopora*, *Sphaerocodium*, and *Girvanella* (Pitcher, 1964). These algae, along with various invertebrate constituents, appear to have been the principal mound builders.

Silurian

Girvanella and *Sphaerocodium*, plus *Ortonella*, are prevalent in Silurian carbonate facies. The earlier Paleozoic problematical blue-green algae *Renalcis*, *Epiphyton*, and *Frutexites* have not been reported from Silurian rocks.

Codiaceans are minor constituents in the Silurian. The Dasycladaceae are represented by taxa similar in kind and number to those in the Ordovician.

Stromatolitic buildups occur locally, but are less widespread than those in earlier Paleozoic strata. Reef facies of Gotland contain locally abundant occurrences of *Solenopora* and *Sphaerocodium* (Hadding, 1950; Manten, 1971).

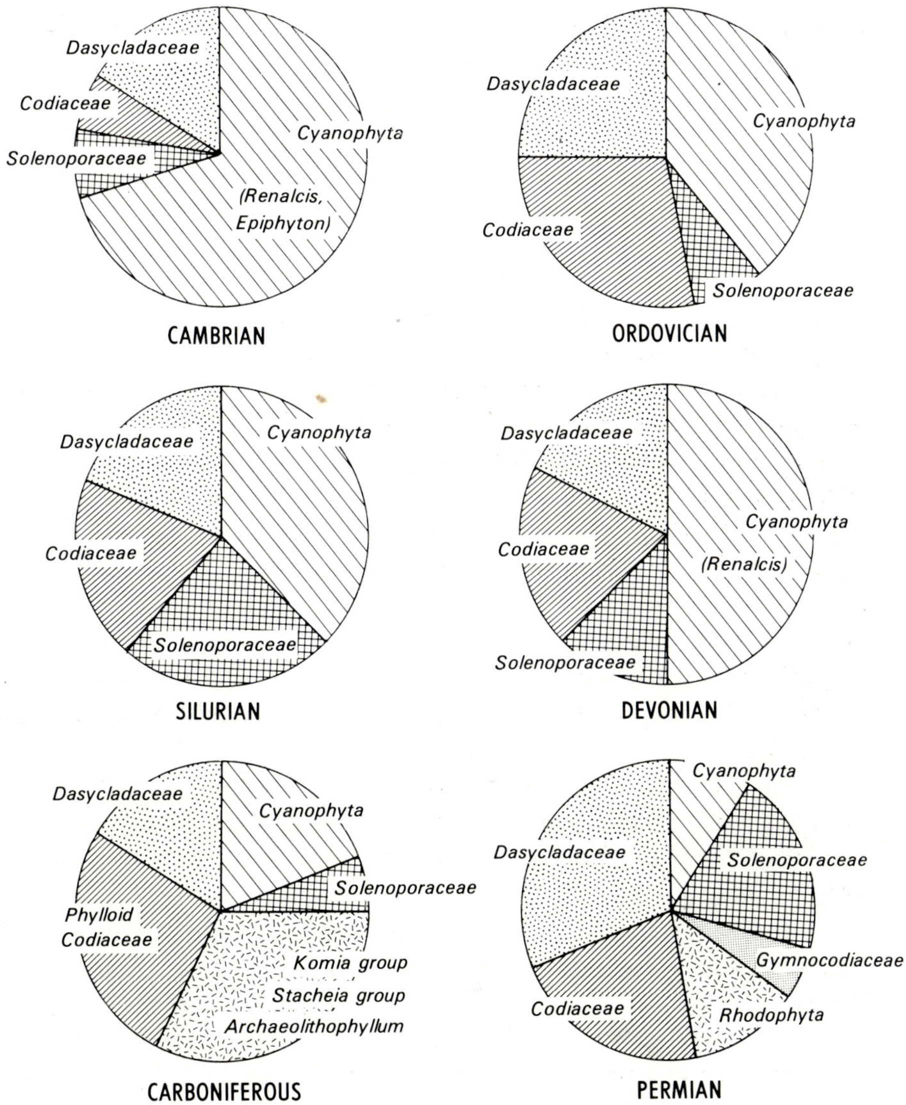

Fig. 161. Estimated assemblage composition of skeletal calcareous algae in Paleozoic periods.

Devonian

Skeletal blue-green algae, including problematical forms, are major contributors to Devonian carbonate facies. *Renalcis* is especially widespread and is a principal constituent along with *Sphaerocodium* in Middle and Upper Devonian stromatoporoid reef facies in Australia and Canada (Wray, 1967; Wray and Playford, 1970).

Solenoporaceae are abundant in some facies associated with reef complexes. Codiaceans and dasycladaceans are relatively unimportant volumetrically, but may be a conspicuous minor element of some carbonate rocks.

Charophytes evolved rapidly in the Devonian and their remains are locally important constituents in nonmarine carbonate facies from this time to the present.

Carboniferous

During the Late Paleozoic various kinds of marine algae contributed substantial amounts of skeletal material to marine carbonate facies. In contrast to the Devonian, most of the algae were not encrusting forms which developed reef frameworks, but were clastic skeletal constituents distributed widely in shelf facies or concentrated in mounds.

Skeletal blue-green algae attained the height of their morphological diversity and filamentous forms are locally abundant. Solenoporaceae are widespread fossils but occur in minor amounts.

In the Early Carboniferous several problematical calcareous red algae belonging to the *Stacheia* group appeared. These forms were succeeded in the Late Carboniferous by the *Ungdarella-Komia* group. Some of these fossils, for example *Komia*, are locally abundant and constitute the principal skeletal constituents of some buildups (Girdley, 1968).

The ancestral coralline *Archaeolithophyllum* appeared near the end of Early Carboniferous time. This encrusting alga was very successful in spreading throughout marine shelf regions of the world during the Late Carboniferous and Early Permian. *Archaeolithophyllum* was capable of erecting a self-supporting skeletal framework, as well as binding and cementing other material, in much the same way as modern corallines. This ancestral coralline, and the related genus *Cuneiphycus* are important framebuilding algae and volumetrically significant skeletal constituents in Upper Carboniferous and Lower Permian carbonate facies in many regions.

Fragmented skeletal remains of the erect, leaflike codiaceans *Eugonophyllum* and *Ivanovia* are abundant in Upper Carboniferous limestones (Wray, 1968). These phylloid codiaceans are the dominant type of skeletal constituent in many Pennsylvanian banks and mounds. Dasycladaceans, such as *Dvinella* and *Epimastopora*, are widespread and may be important in some shelf facies.

Permian

The Upper Permian Capitan reef complex of west Texas and New Mexico is one of the best known examples of an ancient reef. Algae are generally considered to have been one of the major reef-forming biotas in this reef complex, yet identifiable skeletal algae are found in only a few facies. According to Babcock (1975), the principal skeletal algae are *Parachaetetes*, *Hedstroemia*, and *Eugonophyllum*. Some of these encrusting forms are framebuilders locally, however, the most important frameworks volumetrically are formed by sponges, bryozoans, and hydrozoans. A Permian reef complex in southern Tunisia (Boyd et al., 1975) contains similar encrusting red and phylloid algae.

Dasycladacean green algae are often widespread and abundant in Permian shelf carbonate facies, in addition to codiacean and gymnocodiacean red algae. In terms of biological diversity, the calcareous green algae are the most important group during Permian time.

Triassic

Triassic carbonate buildups are well developed in the Calcareous Alps of Austria and adjacent regions of Germany and Italy. Solenoporaceae, mainly *Solenopora*, and a diverse assemblage of dasycladaceans are found in these rocks (Zankl, 1969; Ott, 1972). The solenoporaceans are generally associated with reef facies, whereas the dasycladaceans are widespread in shelf deposits. According to Ott (1972a), the Alpine Triassic may be subdivided into four zones on the basis of different dasycladacean floras.

Codiacean and skeletal blue-green algae appear to be minor elements in Triassic carbonate facies, although *Sphaerocodium* and *Garwoodia* occur in rocks of this age.

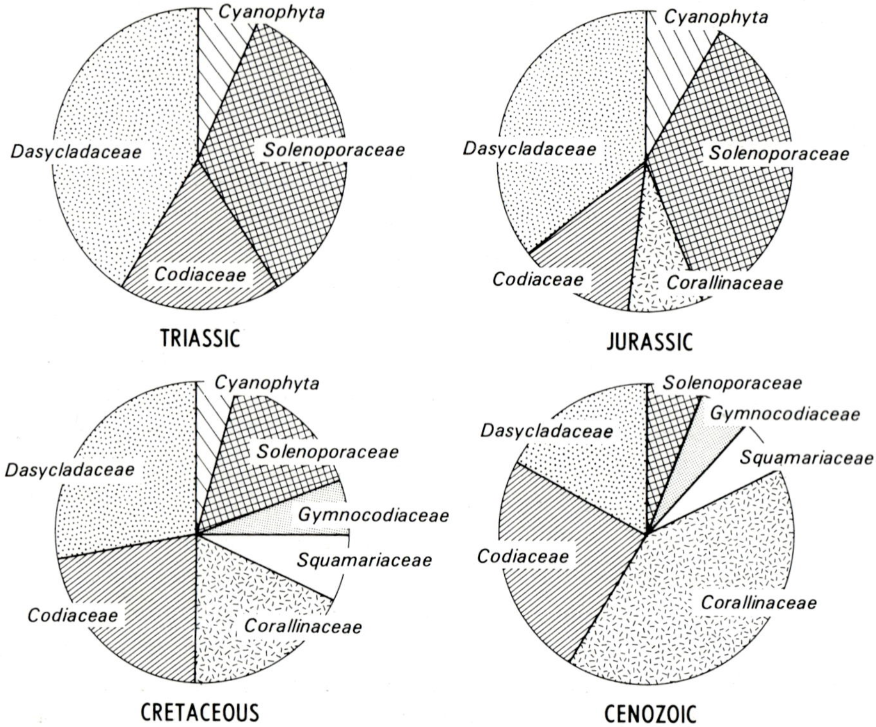

Fig. 162. Estimated assemblage composition of skeletal calcareous algae in the Mesozoic and Cenozoic.

Jurassic

Jurassic time was significant in the evolution of skeletal calcareous algae, because the Solenoporaceae and Dasycladaceae reached the height of their development and modern types of Codiaceae and Corallinaceae first appeared (Johnson, 1964b).

Both the Solenoporaceae and Dasycladaceae occur commonly in Jurassic rocks and the latter group is particularly abundant and widespread in shelf deposits. Codiaceans and corallines are minor elements in Jurassic facies. The filamentous genus *Cayeuxia* is abundant locally. Laminated encrustations, presumably formed by blue-green algae, are significant in the Upper Jurassic sponge mounds of southwestern Germany (Gwinner, 1971).

Cretaceous

Widespread carbonate deposition during Cretaceous time, including reef developments, was favorable for the growth and accumulation of skeletal calcareous algae.

Skeletal blue-green algae are essentially absent in Cretaceous rocks, except for occasional remains of *Girvanella* and *Cayeuxia*. The Cretaceous was the last appearance of the skeletal blue-green algal types so abundant in earlier periods, especially during the Paleozoic era.

Calcareous red algae include the Corallinaceae, Solenoporaceae, a reappearance of the Gymnocodiaceae, and the first appearance of the Squamariaceae. Representatives of all of these families may occur commonly.

Articulated corallines appeared in middle Cretaceous time to join the crustose corallines, as major elements of the marine flora. Although corallines are relatively rare constituents in Lower Cretaceous rocks, they become increasingly important in the Upper Cretaceous where they contribute to reefs and associated facies.

The Codiaceae and Dasycladaceae may be dominant skeletal algae in some Cretaceous carbonate facies, being widespread in shallow-shelf environments. Typically, Upper Cretaceous shelf facies may contain Codiaceae, Dasycladaceae, and Gymnocodiaceae, whereas reefs and other carbonate buildups are characterized by crustose corallines and Squamariaceae.

Cenozoic

Encrusting coralline algae have served as binding and cementing agents in coral reefs throughout the Cenozoic. Unattached crustose corallines and articulated corallines have accumulated in bank proportions.

Paleocene reef facies often contain crustose corallines, Solenoporaceae, and Squamariaceae, whereas shelf environments are characterized by codiaceans (*Halimeda* and *Ovulites*) and various dasycladaceans. Solenoporaceans decreased markedly in abundance at the beginning of Eocene time.

Marine carbonate facies in the latter part of the Cenozoic epic contain an increasing number of coralline genera; however, the skeletal green algae remain more or less unchanged in terms of diversity and abundance. From the mid-Cenozoic to the present, assemblages of calcareous algae are essentially the same as living populations.

PETROLEUM RESERVOIRS AND ALGAL FACIES

Devonian of western Canada

There are three main periods of Devonian carbonate developments containing oil in Alberta. Stromatoporoids are the principal reef framebuilders in these carbonate complexes, but calcareous algae are important constituents in the Late Devonian (Frasnian) Leduc reefs. Reef facies of the Redwater Field in central Alberta contain abundant *Renalcis* (Fig. 163) along with tabular stromatoporoids through intervals of porous and permeable carbonates several tens of meters thick (Wray and Playford, 1970).

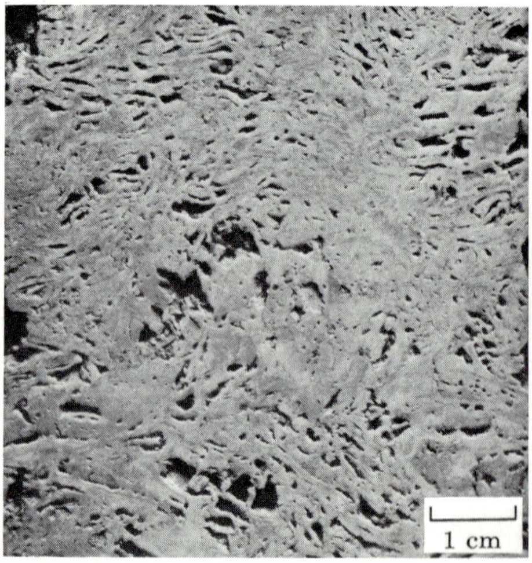

Fig. 163. *Renalcis* facies with tabular stromatoporoids in Devonian Leduc Formation, Redwater Field, Alberta. Reflected light, polished surface.

Fig. 164. Phylloid algal limestone composed of *Ivanovia*. Reservoir facies of Pennsylvanian age, Utah. Reflected light, polished surface.

The occurrence of *Renalcis* at Redwater is unique, because it is one of the few undolomitized Leduc reef complexes. This framebuilding alga may have been similarly widespread and important in other Leduc carbonate complexes where it is now no longer recognizable because of dolomitization. This Upper Devonian algal-stromatoporoid reef framework in Canada is nearly identical to reef facies in the outcropping Devonian reef complexes of Western Australia.

Pennsylvanian of southwestern U.S.

Late Paleozoic phylloid algal limestones in the United States are important petroleum reservoirs. Examples are known from the Paradox Basin (Fig. 165) of southeastern Utah and southwestern Colorado (Choquette and Traut, 1963), and in west Texas and New Mexico. Production is from buildups or lenses of bioclastic limestone composed mainly of the phylloid alga *Ivanovia* (Fig. 164).

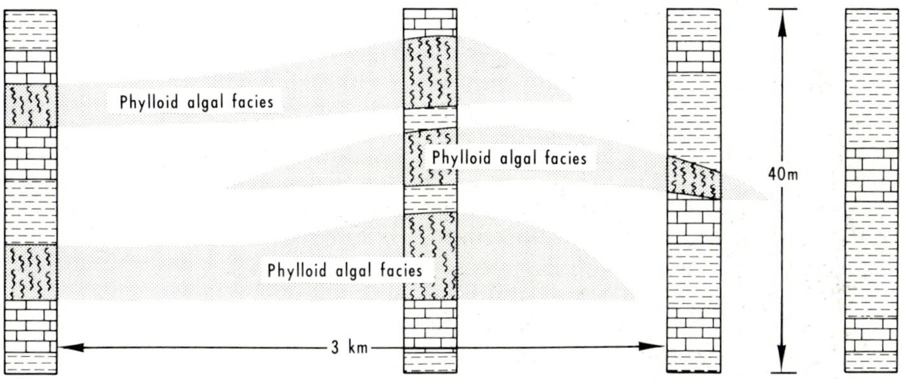

Fig. 165. Diagrammatic cross section of portion of Ismay Field, Colorado and Utah. Production from Upper Carboniferous (Pennsylvanian) carbonate buildups of phylloid algae stacked one above another in three intervals. (After Choquette and Traut, 1963).

Well control in the Ismay Field and studies of nearby outcrop sections (Pray and Wray, 1963) indicate that these buildups are biostromal, flat-bottomed, mound-like deposits up to several hundred meters across, as much as 3 km long, and about 10 m thick. Their geometry and internal characteristics suggest accumulation as local, low-relief, barlike carbonate banks. Occurring in cyclical sequences, nonalgal carbonate rocks and shales deposited in relatively shallow-water environments are associated with *Ivanovia* facies.

Oil production comes from porous and permeable zones within the algal buildups. Porosity in these phylloid algal limestones is both primary, in the sense that the present porosity occupies the site of original pore space between algal fragments, and secondary, having been produced by solution leaching of algal fragments. In both cases, the depositional texture of this phylloid algal sediment is a principal factor in the localization of effective porosity. Porosities up to 26 percent and permeabilities as high as about 900 millidarcies have been reported from the *Ivanovia* facies.

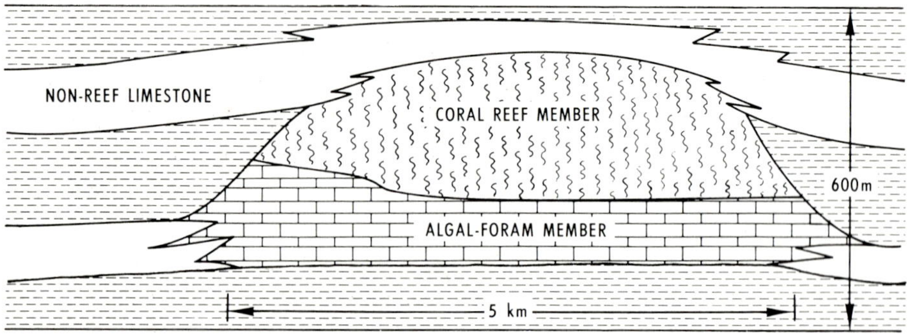

Fig. 166. Diagrammatic cross section of Paleocene reef (Intisar A Field; formerly named Idris), Sirte Basin, Libya. (After Terry and Williams, 1969).

Paleocene of Libya

Major oil fields producing from Cretaceous and Lower Cenozoic carbonates occur in the Sirte Basin of Libya. Reefs and other carbonate accumulations of Paleocene age are the principal oil and gas reservoirs.

The Intisar A reef (Terry and Williams, 1969), a 400-meter-thick bioherm composed mainly of corals and coralline algae (Fig. 166), is a highly productive oil field. This carbonate accumulation began as a bank of large benthonic foraminifera and coralline algae. Subsequently, sedimentation changed to a predominantly coral and coralline algae boundstone. Porosities are greatest in the algal-foraminiferal member, averaging about 27 percent, and are between 20-25 percent throughout much of the overlying coral-algae reef limestone.

The Intisar reef was deposited within a part of a larger shelf province favorable for reef growth, and the site of the accumulation was apparently controlled by a local upwarp in the sea floor that localized growth of coral and algae.

A variety of Lower Cenozoic carbonate facies in Libya are porous and permeable. Reservoir facies occur in coralline algae boundstones (Fig. 167) and in shelf facies composed of dasycladacean algae and pellets (Fig. 168).

ORE DISTRIBUTION AND ALGAL FACIES

Precambrian stromatolites and mineralization

The association of Precambrian stromatolites and ore distribution has been observed in several regions. Mendelsohn (1973) suggested that stromatolitic blue-green algae created an appropriate environment for simultaneous deposition of sediment and copper, leading to the formation of syngenetic stratiform ore deposits in sedimentary rocks. Freshwater blue-green algae are able to survive in water containing as much as 25 ppm copper in solution. Under these conditions the algae could accumulate an important amount of copper, as much as 0.25 percent of their own dry weight. Consequently, these algae growing as mats, which trap and bind particulate sediment, could also have concentrated significant amounts of metal.

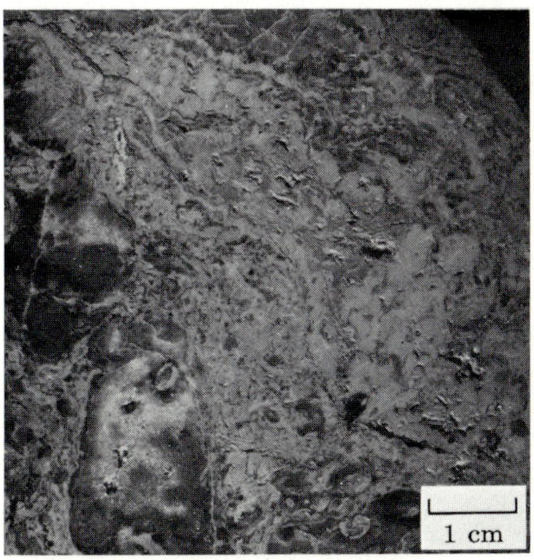

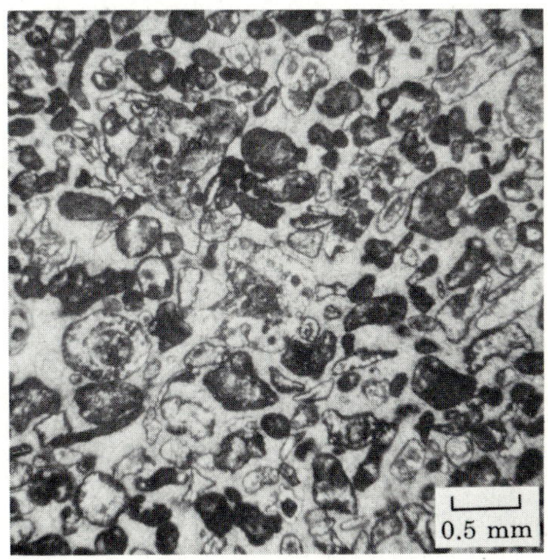

Fig. 167. Coralline algae boundstone. Paleocene, Libya. Reflected light, polished surface.

Fig. 168. Dasycladacean algae and pellet wackestone. Paleocene, Libya. Transmitted light, thin section.

Precambrian copper deposits in Zambia (Northern Rhodesia) are associated with stromatolitic algal developments (Malan, 1964; Garlick and Fleischer, 1972); however, the copper ores are antipathetic to the stromatolites, that is, the algal structures are nearly barren of copper, whereas the associated argillaceous beds are preferentially mineralized (Fig. 169). The carbonate-rich stromatolites, occurring on topographic basement highs, would seem to afford a more favorable chemical environment for metasomatic processes than argillaceous sediments, but other relationships of paleotopography and sedimentary facies were apparently overriding controls.

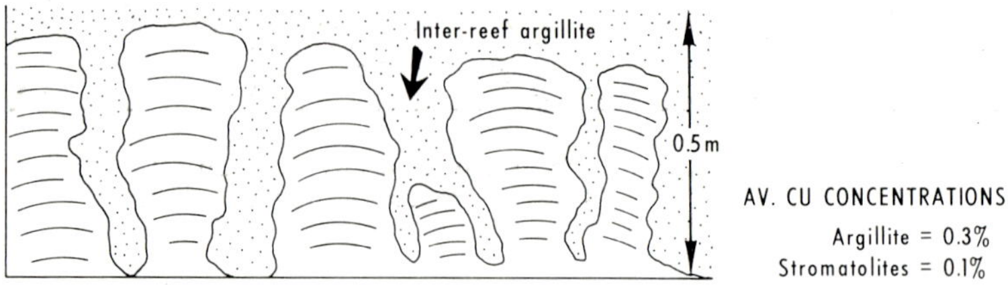

Fig. 169. Distribution of copper in Precambrian stromatolitic reefs, Zambia. Copper mineralization is concentrated in argillaceous facies. (After Malan, 1964).

Regardless of the origin of the ores, the distribution of stromatolitic facies in the Precambrian Zambian copper belt has a direct bearing on their exploration and production. The recognition of stromatolites can be an important part of understanding sedimentary controls and economic considerations in various kinds of stratiform sulfide deposits containing these biosedimentary structures.

Missouri lead district

Located in the central part of the United States, the southeast Missouri lead-zinc-copper district is one of the world's largest. In this region, mineralization in the Bonneterre Formation of Late Cambrian age is spatially related to a barrier reef and associated carbonate facies.

According to Gerdemann and Myers (1972) the depositional facies of these ore-bearing rocks are similar to Recent carbonate bank models (Fig. 170). Algal reefs, composed dominantly of digitate stromatolites, were formed near the shoreline and often developed directly on Precambrian highs. Generally this facies has been completely dolomitized.

The offshore facies, deposited in more open marine conditions, is composed of predominantly fine-grained carbonates, commonly limestones. The back-reef facies is entirely dolomite, obliterating all original sedimentary features, but it appears to have been deposited on a tidal flat.

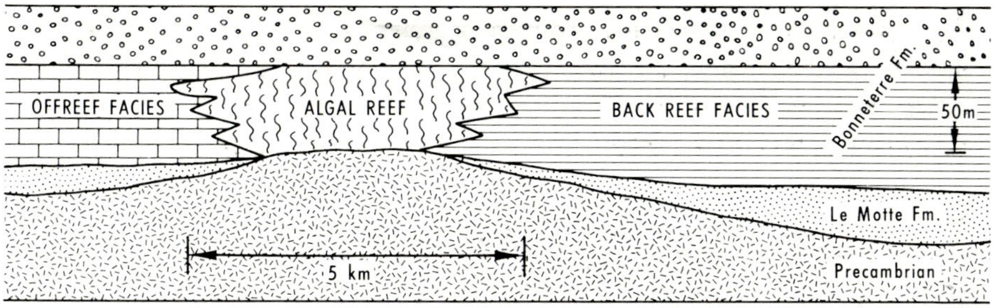

Fig. 170. Diagrammatic cross section of Upper Cambrian mineralized facies, Southeast Missouri Lead District. (After Gerdemann and Myers, 1972).

There is a close relationship between mineralization and regional sedimentary facies patterns, with algal reefs and associated sediments localizing epigenetic ore deposits. The stromatolitic algal reef facies provided a preferential host for mineralization, and Gerdemann and Myers (1972) suggest that biogenically reduced sulfur related to organic material in the algal reefs and other organic-rich sediments precipitated the metals as sulfides.

GLOSSARY

Antheridium -- male gamentangium or sperm-bearing organ.

Apex, apical -- the tip or pointed end of a body or object.

Articulate -- consisting of segments united by joints; jointed.

Articulated corallines -- coralline algae having segmented thalli.

Asexual -- reproduction not involving the union of gametes.

Aspondyle -- irregular arrangement of branches of dasycladacean algae.

Autotrophic -- pertaining to a plant that manufactures its own food.

Axial -- referring to the morphological axis, as distinct from laterial parts.

Bathyal -- the sea floor on the continental slopes.

Benthonic -- bottom-dwelling aquatic plants (and animals); nonplanktonic.

Binomial (binary nomenclature) -- two-named; in biology referring to the genus to which an organism belongs and its own species name.

Blade -- a more or less broad, flattened, foliate part of an erect alga.

Cell -- a basic structural and physiological unit of organisms capable alone or interacting with other cells of performing fundamental functions of life.

Charophyte -- referring to algae or algal organs belonging to the class Charophyta.

Chlorophyll -- green pigment; important in absorption of light in photosynthesis.

Chloroplast -- specialized cytoplasmic body containing chlorophyll.

Coccoid -- referring to round or subspherical cells, usually occurring free from one another within mucilage.

Coccolith -- a calcareous skeletal element of marine planktonic algae (Coccolithophyceae).

Conceptacle -- a cavity in which reproductive organs, such as sporangia, are developed.

Coralline -- referring to algae belonging to the family Corallinaceae (Rhodophyta).

Cortex, cortical -- cells or tissue external to a central axis or inner region.

Crustose -- in the form of a crust; in the case of algae, generally referring to thalli growing flattened against the substrate.

Cytology, cytological -- the science dealing with the cell.

Cytoplasm -- all the protoplasm of a cell outside the nucleus.

Dextral, dextrally -- of or to the right.

Dichotomous, dichotomously -- branching in a forked fashion to form two branches which are usually equal.

Endolithic -- living within rock.

Epiphyte, epiphytic -- a plant that grows upon another plant, but not parasitic.

Epithallium -- a thin surface layer of uncalcified cells in coralline algae.

Eukaryotes, eukaryotic -- morphologically advanced organisms characterized by having the cellular organelles, including the nucleus, bounded by membranes.

Euspondyle -- a regular arrangement of primary branches into whorls (Dasycladaceae).

Family -- in plant taxonomy, a group of genera; families are grouped into orders.

Filament, filamentous -- a type of thallus consisting of one or more slender rows or lines of closely adjoined cells, with or without a mucilaginous sheath.

Foliate -- shaped like a leaf.

Gamentangium -- an organ producing gametes.

Gamete -- a cell that fuses with another in sexual reproductive processes.

Geniculum -- an uncalcified joint between segments of an articulated coralline alga.

Genus -- a group of structurally or phylogenetically related species.

Golgi apparatus -- specialized vesicles in the cytoplasm, associated with various organelles of the cell.

Gyrogonites -- a small calcified part of the oogonium (Charophyta).

Heterocyst -- a cell larger than, or different from neighboring cells.
Holdfast -- a basal attachment device of an alga.
Hypothallium, hypothallial -- the basal-most tissue of Corallinaceae in which the filaments are oriented parallel to the substrate. Coaxial hypothallium is hypothallial tissue in which cells are arranged in arched rows and occupy an inner (medullary) position.

Intercellular -- lying between cells.
Intracellular -- lying within cells.
Intergeniculum -- a calcified segment between uncalcified joints (genicula) in articulated coralline algae.
Intertidal -- the coastal zone lying within the upper level of high tide and the lowest level of low tide; the zone that is alternately covered and uncovered by tidal action; littoral.

Medulla, medullary -- the inner or central region of a thallus.
Meristem -- region or cell layer in which most of the vegetative cell division occurs.
Metabolism, metabolic -- the process by which nutritive matter is built into living material.
Metaspondyle -- arrangement of primary branches into clusters of three or six occurring in regular whorls (Dasycladaceae).
Monostromatic -- having cells in a single layer.

Oogonium -- a single-celled female reproductive organ (Charophyta).
Organelle -- a membrane-bounded, specialized region within a cell.

Perithallium, perithallial -- cellular tissue of Corallinaceae lying between the hypothallium and the laterial meristem and made up of filaments oriented essentially perpendicular to the substrate.
Photosynthesis -- a process in which carbon dioxide and water are chemically combined to form carbohydrate, the energy for the process being sunlight.
Phylloid -- resembling a leaf; leaflike.
Phylum -- a primary division of the plant or animal kingdom.
Physiology -- science of the functions and activities of organisms.

Phytoplankton -- free-floating plant life, usually microscopic.

Pinnate -- having divisions on each side of an axis; featherlike.

Pit connection -- a porelike opening in the cell wall of the Florideae (Rhodophyta) through which cytoplasmic material pass. A primary pit connects cells in the same filament; a secondary pit connects adjacent cells not belonging to the same filament.

Plankton -- free-floating plant and animal life.

Plastid -- a cellular organelle in which carbohydrate metabolism is localized.

Prokaryotes, prokaryotic -- morphologically primitive organisms (bacteria and blue-green algae; mostly single cells or simple filaments) which do not have DNA separated from the cytoplasm by an envelope.

Protoplasm -- living substance.

Rhizoid -- a rootlike attachment filament.

Segment -- a division of a jointed, segmented, or divided thallus.

Sinistral -- of or to the left.

Siphonaceous -- possessing a tubular structure (Chlorophyta).

Sori -- a group of sporangial compartments.

Species -- a group of individuals having common attributes.

Sporangium -- a spore-producing structure; spore case.

Spore -- a reproductive cell that develops into a plant without a union with other cells.

Stipe -- the stemlike, basal portion of a thallus.

Subcortex, subcortical -- beneath or within the cortex; generally a tissue between cortex and medulla.

Sublittoral -- the sea floor on continental shelves.

Taxon -- a term for any taxonomic rank.

Taxonomy -- the science dealing with the describing, naming, and classification of organisms.

Thallus -- the entire plant body of an alga (without true roots, stem, or leaves).

Tissue -- a group of cells of similar structure that performs a specialized function.

Trichome -- a thread of cells in filamentous cyanophytes, but without the sheath.

Unicell -- an organism consisting of a single cell.

Utricle -- swollen, saclike branches or filaments within a thallus.

Vacuole -- a specialized region of a cell.

Whorl -- a circle of branches arising from a stem, usually equally spaced.

Zoospore -- a motile (moving) spore.

REFERENCES

Adey, W. H., 1966. The distribution of saxicolous crustose corallines in the northwestern North Atlantic. J. Phycol., 2:49-54.

Adey, W. H., 1970. The effects of light and temperature on growth rates in boreal-subarctic crustose corallines. J. Phycol., 6:269-276.

Adey, W. H., 1970a. A revision of the Foslie crustose coralline herbarium. Norske Vidensk-Selsk. Skr. Kl., 1:1-46.

Adey, W. H., 1970b. Some relationships between crustose corallines and their substrate. Sci. Islandica, 2:21-25.

Adey, W. H., 1975. The algal ridges and coral reefs of St. Croix: their structure and Holocene development. Atoll Research Bull., 187:1-67.

Adey, W. H. and Adey, P. J., 1973. Studies of the biosystematics and ecology of the epilithic crustose Corallinaceae of the British Islands. British Phycol. Jour., 8:343-407.

Adey, W. H. and Boykins, W. T., (in press). The crustose coralline algae of the Hawaiian Archipelago. Smithsonian Inst. Contrib. Mar. Sci.

Adey, W. H. and Burke, R., 1976. Holocene bioherms (algal ridges and bank-barrier reefs) of the eastern Caribbean. Geol. Soc. Am. Bull., 87:95-109.

Adey, W. H. and Johansen, H. W., 1972. Morphology and taxonomy of Corallinaceae with special reference to *Clathromorphum, Mesophyllum,* and *Neopolyporolithon* gen nov. (Rhodophyceae, Cryptonemiales). Phycologia, 11(2): 159-180.

Adey, W. H. and Macintyre, I. G., 1973. Crustose coralline algae: a re-evaluation in the geological sciences. Geol. Soc. Am. Bull., 84:883-904.

Adey, W. H. and McKibbon, D. L., 1970. Studies of the maerl species *Phymatolithon calcareum* (Pallas) nov. comb. and *Lithothamnium coralloides* Crouan in the Ria de Vigo. Bot. Marina, 13:100-106.

Adey, W. H. and Vassar, J. M., 1975. Colonization, succession and growth rates of tropical crustose coralline algae (Rhodophyta, Cryptonemiales). Phycologia, 14(2):55-69.

Ahr, W. M., 1971. Paleoenvironment, algal structures, and fossil algae in the Upper Cambrian of central Texas. J. Sediment. Petrol., 41:205-216.

Aitken, J. D., 1967. Classification and environmental significance of cryptalgal limestones and dolomites with illustrations from the Cambrian and Ordovician of southwestern Alberta. J. Sediment. Petrol., 37:1163-1178.

Arnott, H. J. and Pautard, F. G. E., 1970. Calcification in plants. In: H. Schraer (Editor), Biological Calcification: Cellular and Molecular Aspects. Appleton-Century-Crofts, New York, 375-446.

Baas-Becking, L. G. M. and Galliher, E. W., 1931. Wall structure and mineralization in coralline algae. J. Phys. Chem., 35:467-479.

Babcock, J. A., 1974. The role of algae in the formation of the Capitan Limestone (Permian, Guadalupian), Guadalupe Mountains, West Texas-New Mexico. Thesis, Univ. Wisconsin, 241 pp.

Bailey, A. and Bisalputra, T., 1970. A preliminary account of the application of thin-sectioning, freeze-etching, and scanning electron microscopy to the study of coralline algae. Phycologia, 9:83-101.

Barghoorn, E. S. and Tyler, S. A., 1965. Microorganisms from the Gunflint Chert. Science, 147:563-577.

Barton, E. S., 1901. The genus *Halimeda*. Siboga-Exped. Monogr. 60. E. J. Brill, Leiden, 32 pp.
Bassoulet, J.-P., Bernier, P., Deloffre, R., Genot, P., Jeffrezo, M., Poignant, A.-F. and Segonzac, G., 1975. Réflexions sur la systématique des Dasycladales fossiles - Étude critique de la terminologie et importance relative des critères de classification. Géobios, 8(4):259-290.
Berryhill, H. L., Schweinfurth, S. P. and Kent, B. H., 1971. Coal-bearing Upper Pennsylvanian and Lower Permian rocks, Washington area, Pennsylvania. U. S. Geol. Surv. Prof. Pap. 621, 47 pp.
Bertrand-Sarfati, J., 1972. Stromatolites columnaires du Précambrian supérieur du Sahara nord-occidental. Centre Rech. Zones Arid., Sér. Géol., 14:245 pp.
Black, M., 1933. The algal sediments of Andros Island, Bahamas. Phil. Trans. Roy. Soc. London, ser. B, 222:165-192.
Böhm, E. L., 1973. Studies on the mineral content of calcareous algae. Bull. Mar. Sic., 23:177-190.
Bold, H. C., 1967. Morphology of plants. Harper and Row, New York, 541 pp.
Bornemann, J. G., 1886. Die Versteinerungen des Dambrischen Schichten-Systems der Insel Sardinien nebst vergleichenden Untersuchungen über anologe Vorkommnisse aus andern Ländern. Erst. Abt. Ksl. Leop.-Carol. Deut. Akad. Naturforsch. 51:147 pp.
Bosellini, A. and Ginsburg, R. N., 1971. Form and internal structure of Recent algal nodules (rhodolites) from Bermuda. J. Geol., 79:669-682.
Bosence, D. W. J., 1976. Ecological studies on two unattached coralline algae from Western Ireland. Palaeontology, 19:365-395.
Boyd, D. W., Driggs, A., Newell, N. D., Rigby, J. K. and Stehli, F. G., 1975. Permian reef complex, Tunisia (abs.). Geol. Soc. Am. Abstr. Prog., 7(7): 1006.
Byrnes, J. G., 1968. Notes on the nature and environmental significance of the Receptaculitaceae. Lethaia, 1:368-381.

Cabioch, J., 1972. Etude sur les Corallinacees. II. La Morphogenese; Consequences systematiques et phylogenetiques. Cahiers de Biologie Marine, 8:137-288.
Chave, K. E., 1952. A solid solution between calcite and dolomite. J. Geol., 60:190-192.
Chave, K. E., Deffeyes, K. S., Weyl, P. K., Garrels, R. M. and Thompson, M. E., 1962. Observations on the solubility of skeletal carbonates in aqueous solutions. Science, 137:33-34.
Chave, K. E. and Wheeler, B. D., 1965. Mineralogic changes during growth in the red alga, *Clathromorphum compactum*. Science, 147:621.
Choquette, P. W. and Traut, J. D., 1963. Pennsylvanian carbonate reservoirs, Ismay Field, Utah and Colorado. In: R. O. Bass (Editor), Shelf Carbonates of the Paradox Basin. Four Corners Geol. Soc. Sympos., 4th Field Conf., 157-184.
Chuvashov, B. I., 1965. *Katavella*, new genus of fossil red algae. Paleontol. Zhur., 1965(2):144-146.
Chuvashov, B. I., 1971. A new genus of Late Paleozoic red algae. Paleontol. Zhur., 1971(2):85-89.
Cloud, P. E., 1942. Notes on stromatolites. Am. J. Sci., 240:363-379.
Cloud, P. E. and Semikhatov, M. A., 1969. Proterozoic stromatolite zonation. Am. J. Sci., 267:1017-1061.
Colinvaux, L. H., Wilbur, K. M. and Watabe, N., 1965. Tropical marine algae: growth in laboratory culture. J. Phycol., 1(2):69-78.
Coron, C. R. and Textoris, D. A., 1974. Non-calcareous algae in Silurian carbonate mud mound, Indiana. J. Sediment. Petrol., 44:1248-1250.
Croley, F. C. and Dawes, C. J., 1970. Ecology of the algae of a Florida Key. I. a preliminary checklist, zonation, and seasonality. Bull. Mar. Sci., 20:165-185.

Dahl, A. L., 1973. Surface area in ecological analysis: quantification of benthic coral-reef algae. Mar. Biol., 23:239-249.
Dalrymple, D. W., 1965. Calcium carbonate deposition associated with blue-green algal mats, Baffin Bay, Texas. Texas Inst. Mar. Sci., 10:187-200.
d'Archiac, E. J., 1843. Description géologique du département de l'Aisne. Soc. géol. France Mém., 5(2):129-418.
Dawson, E. Y., 1966. Marine botany: an introduction. Holt, Rinehart and Winston, New York, 371 pp.
Deelman, J. C., 1972. On mechanisms causing birdseye structures. N. Jb. Geol. Paläont. Mh., 1972(10):582-595.
Denizot, M. and Massieux, M., 1965. Présence de *Peyssonnelia antiqua* dans le calcaire "Ypréso-lutétien" de la montagne d'Alaric. Rev. Micropaléontol., 8(2):96-102.
Dixon, P. S., 1970. A critique of the taxonomy of marine algae. Ann. N. Y. Acad. Sci., 175:617-622.
Donaldson, J. A., 1963. Stromatolites in the Denault Formation, Marion Lake, Coast of Labrador, Newfoundland. Geol. Surv. Can. Bull., 102:33 pp.
Drouet, F., 1968. Revision of the classification of the Oscillatoriaceae. Acad. Nat. Sci. Phila. Mon., 15:1-370.
Drouet, F., 1973. Revision of the Nostocaceae with cylindrical trichomes. Hafner Press, New York, 292 pp.

Elliott, G. F., 1955. The Permian calcareous alga *Gymnocodium*. Micropaleontology, 1:83-90.
Elliott, G. F., 1957. New calcareous algae from the Arabian Peninsula. Micropaleontology, 3:227-230.
Elliott, G. F., 1958. Algal debris-facies in the Cretaceous of the Middle East. Palaeontology, 1:254-259.
Elliott, G. F., 1960. Fossil calcareous algal floras of the Middle East. Geol. Soc. London Quart. Jour., 115:217-232.
Elliott, G. F., 1961. The sexual organization of Cretaceous *Permocalculus* (calcareous algae). Palaeontology, 4:82-84.
Elliott, G. F., 1963. A Liassic *Pycnoporidium* (calcareous algae). Eclog. Geol. Helv., 56(1):179-181.
Elliott, G. F., 1965. The interrelationships of some Cretaceous Codiaceae (calcareous algae). Palaeontology, 8:199-203.
Elliott, G. F., 1965a. Tertiary solenoporacean algae and the reproductive structures of the Solenoporaceae. Palaeontology, 7:695-702.
Elliott, G. F., 1968. Permian to Palaeocene calcareous algae (Dasycladaceae) of the Middle East. Brit. Mus. (Nat. Hist.) Bull., Geol. Suppl. 4:111 pp.
Elliott, G. F., 1970. Calcareous algae new to the British Carboniferous. Palaeontology, 13:443-450.
Elliott, G. F., 1973. A Miocene solenoporoid alga showing reproductive structures. Palaeontology, 16:223-230.
Ellis, J., 1755. An Essay towards a Natural History of the Corallines, and other Marine Productions of the like kind, commonly found on the Coasts of Great Britain and Ireland. London, 103 pp.
Emery, K. O., Tracey, J. I. and Ladd, H. S., 1954. Geology of Bikini and nearby atolls. Pt. 1, Geology. U. S. Geol. Surv. Prof. Pap. 260-A, 265 pp.
Endo, R., 1951. Stratigraphical and paleontological studies of the later Paleozoic calcareous algae in Japan; I. Several new genera from the Sakamotozawa section, Hikoroichi-mura, Kesen-gun, in the Kitakami Mountain land. Trans. Proc. Palaeont. Soc. Japan, n. S., 4:121-129.
Endo, R., 1959. Stratigraphical and paleontological studies of the Later Paleozoic calcareous algae in Japan, XIV: Fossil algae from the Nyugawa Valley in the Hida Massif. Saitama Univ. Sic. Rept. Ser. B. 3:177-217.

Endo, R., 1961. Phylogenetic relationships among the calcareous algae. Sci. Repts. Saitama Univ. Series B, 52 pp.
Esteban, M., 1974. Caliche textures and "Microcodium". Bull. Soc. Geol. It., 92:105-125.

Fairchild, T. R., Schopf, J. W. and Folk, R. L., 1973. Filamentous algal microfossils from the Caballos Novaculite, Devonian of Texas. J. Paleontol., 47:946-952.
Fisher, T. R., 1975. Fossil genus *Komia* (Middle Pennsylvanian): morphology, microstructure, possible systematics and occurrence in stratigraphic record (abs.). Am. Assoc. Petroleum Geologists, Ann. Mtgs. Abstr., 2:24.
Flügel, E. and Wolf, K. H., 1969. "Sphaerocodium" (Algen) aus dem Devon von Deutschland, Marokko und Australien. N. Jb. Geol. Paläont. Mh., 1969(2): 88-103.
Fournie, D., 1967. Les Porostromata du Paléozoique. Etude bibliographique. Bull. Centre Rech. PAU-SNPA, 1(1):21-41.
Fritch, F. E., 1935-1945. The structure and reproduction of the algae. University Press, Cambridge, 1 (1935), 791 pp.; 2 (1945), 939 pp.

Garlick, W. G. and Fleischer, V. D., 1972. Sedimentary environment of Zambian copper deposition. Geol. Mijnb., 51:277-298.
Garrett, P., 1970. Phanerozoic stromatolites: noncompetitive ecologic restriction by grazing and burrowing animals. Science, 169:171-173.
Garwood, E. J., 1913. On the important part played by calcareous algae at certain geological horizons, with special reference to the Paleozoic rocks. Geol. Mag. Dec. 5, 10:440-553.
Garwood, E. J., 1931. Important additions to our knowledge of the fossil calcareous algae since 1913, with special reference to the Pre-Cambrian and Palaeozoic rocks. Geol. Soc. London Quart. Jour., 87:48-100.
Gebelein, C. D., 1969. Distribution, morphology, and accretion rate of Recent subtidal algal stromatolites, Bermuda. J. Sediment. Petrol., 39:49-69.
Gerdemann, P. E. and Myers, H. E., 1972. Relationships of carbonate facies patterns to ore distribution and to ore genesis in the southeast Missouri lead district. Econ. Geol., 67:426-433.
Ginsburg, R. N., 1955. Recent stromatolitic sediments from south Florida (abs.). J. Paleontol., 29:723.
Ginsburg, R. N., 1956. Environmental relationships of grain size and constituent particles in some South Florida carbonate sediments. Am. Assoc. Petroleum Geologists Bull., 40:2384-2427.
Ginsburg, R. N., 1960. Ancient analogues of Recent stromatolites. Internat. Geol. Cong., 21st, Copenhagen, Pt. 22:26-35.
Ginsburg, R. N. and James, N. P., 1973. British Honduras by submarine. Geotimes, 18:23-24.
Ginsburg, R. N. and James, N. P., 1974. Spectrum of reef building communities in the western Atlantic. In: A. M. Ziegler (Editor), Principles of Benthic Community Analysis. Comparative Sediment. Lab., Univ. Miami, 7.1-7.22.
Ginsburg, R. N. and James, N. P., 1976. Submarine botryoidal aragonite in Holocene reef limestones, Belize. Geology, 4:431-436.
Ginsburg, R. N., Rezak, R. and Wray, J. L., 1971. Geology of calcareous algae (Notes for a short course). Comparative Sediment. Lab., Univ. Miami, 64 pp.
Girdley, W. A., 1968. *Komia* banks (Pennsylvanian) of southwestern San Juan Mountains, Colorado (abs.). Am. Assoc. Petroleum Geologists Bull., 52:529.
Gleason, P. J., 1972. The origin, sedimentation and stratigraphy of a calcitic mud located in the southern freshwater Everglades. Thesis, Pennsylvania State Univ., 355 pp.

Glück, H., 1912. Eine neue gesteinbildende Siphonee (Codiacée) aus dem marinen Tertiar von Sud-deutschland. Mitt. badische Geol. Landesanst. Bd., 4:3-24.
Goldring, W., 1938. Algal barrier reef in the lower Ozarkian of New York with a chapter on the importance of coralline algae as reef builders through the ages. New York State Mus. Bull., 315:5-75.
Golubic, S., Perkins, R. D. and Lukas, K. J., 1975. Boring microorganisms and microborings in carbonate substrates. In: R. W. Frey (Editor), The Study of Trace Fossils, a Synthesis of Principles, Problems, and Procedures in Ichnology. Springer-Verlag, New York, 229-259.
Goreau, T. F., 1963. Calcium carbonate deposition by coralline algae and corals in relation to their roles as reef-builders. Ann. N. Y. Acad. Sci., 109:127-167.
Goreau, T. F. and Goreau, N. I., 1973. The ecology of Jamaican coral reefs II; Geomorphology, zonation and sedimentary phases. Bull. Marine Sci., 23:399-464.
Grambast, L. J., 1972. Principes de l'utilisation stratigraphique des Charophytes: Applications au Paléogène d'Europe occidentale. Fr. Bur. Rech. Geol. Minieres Mem. 77(1):319-328.
Grambast, L. J., 1974. Phylogeny of the Charophyta. Taxon, 23(4):463-481.
Gwinner, M. P., 1971. Carbonate rocks of the Upper Jurassic in SW Germany. In: G. Müller (Editor), Sedimentology of Parts of Central Europe. Guidebook, VIII Internat. Sedimentol. Congr., Heidelberg, 193-207.

Hadding, A., 1950. Silurian reefs of Gotland. J. Geol., 58:402-409.
Hall, J. D., 1883. *Cryptozoon (proliferum)* N. G. (and sp.). New York State Mus. Ann. Rept. 36: pl. 6 and expl.
Heckel, P. H., 1975. Solenoporid red algae (*Parachaetetes*) from upper Pennsylvanian rocks in Kansas, J. Paleontol., 49:662-673.
Herak, M., 1965. Comparative study of some Triassic Dasycladaceae in Yugoslavia. Geol. Vjesnik, Zagreb, 18(1):3-34.
Hillis, L. W., 1959. A revision of the genus *Halimeda* (order Siphonales). Texas Inst. Mar. Sci., 6:321-403.
Høeg, O. A., 1927. *Dimorphosiphon rectangulare,* preliminary note on a new Codiaceae from the Ordovician of Norway. Avh. Norske vidensk.-akad. Oslo, I Mat.-Nat. Kl., 4:1-15.
Hoffman, P., 1967. Algal stromatolites: use in stratigraphic correlation and paleocurrent determination. Science, 157:1043-1045.
Hofmann, H. J., 1969. Attributes of stromatolites. Geol. Surv. Can. Pap. 69-39:58 pp.
Hofmann, H. J., 1973. Stromatolites: characteristics and utility. Earth-Sci. Rev., 9:339-373.
Hofmann, H. J., 1975. Stratiform Precambrian stromatolites, Belcher Islands, Canada: relations between silicified microfossils and microstructure. Am. J. Sci., 275:1121-1132.
Hommersand, M. H., 1972. Algae. In: D. N. Lapedes (Editor), 1972 McGraw-Hill Yearbook of Science and Technology. McGraw-Hill, New York, 106-112.
Honjo, S., 1976. Coccoliths: production, transportation and sedimentation. Mar. Micropaleontol., 1:65-79.
Horn of Rantzien, H., 1959. Recent charophyte fructifications and their relations to fossil charophyte gyrogonites. Kungl. Svenska Vet. Akad. Arkiv Bot., Ser. 2 4(7):165-332.
Horowitz, A. S. and Potter, P. E., 1971. Introductory petrography of fossils. Springer-Verlag, New York, 302 pp.
Hudson, J. D., 1970. Algal limestones with pseudomorphs after gypsum from the Middle Jurassic of Scotland. Lethaia, 3:11-40.
Huvé, M. P., 1954. Étude experimentale de la reinstallation d'um trottoir à *Tenarea* en Méditerranée occidentale. Compt. Rend., 239:323-325.

James, N. P., 1972. Holocene and Pleistocene calcareous crust (caliche) profiles: criteria for subaerial exposure. J. Sediment. Petrol., 42:817-836.

Johansen, H. W., 1969. Morphology and systematics of coralline algae with special reference to *Calliarthron*. Univ. Calif. Publ. Bot., 49:98 pp.

Johnson, J. H., 1943. Geological importance of calcareous algae with annotated bibliography. Colorado School Mines Quart., 38(1):102 pp.

Johnson, J. H., 1954. Cretaceous Dasycladaceae from Gillespie County, Texas. J. Paleontol., 28:787-790.

Johnson, J. H., 1956. *Archaeolithophyllum,* a new genus of Paleozoic coralline algae. J. Paleontol., 39:53-55.

Johnson, J. H., 1957. Bibliography of fossil algae: 1942-1955. Colorado School Mines Quart., 52(2):92 pp.

Johnson, J. H., 1960. Paleozoic Solenoporaceae and related red algae. Colorado School Mines Quart., 55(3):77 pp.

Johnson, J. H., 1961. Limestone-building algae and algal limestones. Colorado School Mines, Golden, 297 pp.

Johnson, J. H., 1961a. Studies of Ordovician algae. Colorado School Mines Quart., 56(2):101 pp.

Johnson, J. H., 1963. Pennsylvanian and Permian algae. Colorado School Mines Quart., 58(3):211 pp.

Johnson, J. H., 1964. Lower Devonian algae and encrusting foraminifera from New South Wales. J. Paleontol., 38:98-108.

Johnson, J. H., 1964a. Paleocene calcareous red algae from northern Iraq. Micropaleontology, 10:207-216.

Johnson, J. H., 1964b. The Jurassic algae. Colorado School Mines Quart., 59(2):129 pp.

Johnson, J. H., 1965. Fossil algae from Guatemala. Colorado School Mines Prof. Contrib., 1:152 pp.

Johnson, J. H., 1966. A review of the Cambrian algae. Colorado School Mines Quart., 61(1):162 pp.

Johnson, J. H., 1967. Bibliography of fossil algae, algal limestones, and the geological work of algae, 1956-1965. Colorado School Mines Quart., 62(4):148 pp.

Johnson, J. H., 1968. Lower Cretaceous algae from Texas. Colorado School Mines Prof. Contrib., 4:71 pp.

Johnson, J. H., 1969. A review of the Lower Cretaceous algae. Colorado School Mines Prof. Contrib., 6:180 pp.

Johnson, J. H. and Adey, W. H., 1965. Studies of *Lithophyllum* and related algal genera. Colorado School Mines Quart., 60(2):105 pp.

Johnson, J. H. and Konishi, K., 1956. A review of Mississippian algae. Colorado School Mines Quart., 51(4):84 pp.

Johnson, J. H. and Konishi, K., 1958. A review of Devonian algae. Colorado School Mines Quart., 53(2):84 pp.

Johnson, J. H. and Konishi, K., 1959. A review of Silurian algae (Gotlandian) algae. Colorado School Mines Quart., 54(1):82 pp.

Jux, U., 1966. *Palaeoporella* im Boda-Kalk von Dalarne. Palaeontographica, B, 118(4/6):153-165.

Kalkowsky, E., 1908. Oolith und Stromatolith im norddeutschen Buntsandstein. Zeitschr. Deutsch. Geol. Ges., 60:68-125.

Kamptner (von), E., 1958. Uber das system und die Stammesgeschichte der Dasycladaceen (Siphonae Verticillatae). Ann. Geol. Paläont. Nat. Mus., Wien, 62:95-122.

Kazmierczak, J., 1975. Colonial Volvocales (Chlorophyta) from the Upper Devonian of Poland and their paleoenvironmental significance. Acta Palaeont. Polon., 20(1):73-85.

Kendall, C. G. St. C. and Skipwith, P. A. d'E., 1968. Recent algal mats of a Persian Gulf lagoon. J. Sediment. Petrol., 38:1040-1058.

Kesling, R. V. and Graham, A., 1962. *Ischadites* is a dasycladacean alga. J. Paleontol., 36:943-952.

Klement, K. W. and Toomey, D. F., 1967. Role of the blue-green alga *Girvanella* in skeletal grain destruction and lime-mud formation in the Lower Ordovician of West Texas. J. Sediment. Petrol., 37:1045-1051.

Kochansky-Devidé, V. and Gusić, I., 1971. Evolutions-Tendenzen der Dasycladaceen mit besonderer Berücksichtigung neuer Funde in Jugoslawien. Paläont. Zeit., 45(1/2):82-91.

Konishi, K., 1954. *Succodium*, a new codiacean genus, and its algal associates in the Late Permian Kuma Formation of southern Kyushu, Japan. J. Fac. Sci., Univ. Tokyo, Sec. II, 9(11):225-240.

Konishi, K., 1961. Studies of Paleozoic Codiaceae and allied algae. Pt. 1: Codiaceae (excluding systematic descriptions). Kanazawa Univ. Sci. Rpts., 7(2):159-261.

Konishi, K. and Epis, R. C., 1962. Some early Cretaceous calcareous algae from Cochise County, Arizona. Micropaleontology, 8:67-76.

Konishi, K. and Wray, J. L., 1961. *Eugonophyllum*, a new Pennsylvanian and Permian algal genus. J. Paleontol., 35:659-666.

Korde, K. B., 1959. The morphology and systematic position of representatives of the genus *Epiphyton*. Acad. Sci. USSR, Doklady, 126(5):1087-1089.

Korde, K. B., 1961. Cambrian algae from southeastern part of Siberian platform. Acad. Sci. USSR, Paleont. Inst. Trudy, 89:147 pp.

Korde, K. B., 1973. Cambrian algae. Acad. Sci. USSR, Paleont. Inst. Trudy, 139:349 pp.

Kotila, D. A., 1973. Algae and paleoecology of algal and related facies, Morrow Formation, northeastern Oklahoma. Thesis, Univ. Oklahoma, 231 pp.

Kozlowski, R. and Kazmierczak, J., 1968. On two Ordovician calcareous algae. Acta Palaeont. Polon., 13(3):325-346.

Kulik, Ye. L., 1964. Beresellids from the Carboniferous of the Russian Platform. Paleont. Zhurn., 1964(2):99-114. [English translation: Internat. Geol. Rev., 7(9):1643-1654].

Kylin, H., 1956. Die Gattungen der Rhodophyceen. CWK Gleerups Förlag, Lund, 673 pp.

Lamarck, J. B., 1816. Histoire Naturelle des Animaux sans Vertèbres. Paris, 2:568 pp.

Lamouroux, J. V. F., 1812. Extrait d'un mémoire sur la classification des polypiers coralligènes non entièrement pierreux. Nouv. Bull. Sci. Soc. Philom. Paris, 3:181-188.

Lamouroux, J. V. F., 1816. Histoire des Polypiers Coralligène Flexibles, Vulgairement Nommés Zoophytes. E. Poisson, Caen, 559 pp.

Lamouroux, J. V. F., 1824. Corallina: or a clasical arrangement of flexible coralline polypidoms, selected from the French of J. V. F. Lamouroux. London, 303 pp.

Land, L. S., 1970. Carbonate mud: production by epibiont growth on *Thalassia testudinum*. J. Sediment. Petrol., 40:1361-1363.

Lang, J. C., 1974. Biological zonation at the base of a reef. Am. Sci., 62:272-281.

Lemoine, M., 1940. Les algues calcaires de la zone neritique. Soc. Biog., 7:75-138.

Levring, T., 1969. Classification of the algae. In: T. Levring, H. A. Hoppe and O. J. Schmid, Marine Algae: A Survey of Research and Utilization. Cram, de Gruyter & Co., Hamburg, 47-125.

Lewin, J. C., 1962. Calcification. In: R. A. Lewin (Editor), Physiology and Biochemistry of Algae. Academic Press, New York, 457-465.

Link, H. F., 1834. Sur les zoophytes en général et en particulier sur certaines plantes qu'on a confondues avec eux. Ann. Sci. Nat. (Bot)., 2:321-331.

Littler, M. M., 1972. The crustose Corallinaceae. Oceanogr. Mar. Biol. Ann. Rev., 10:311-347.

Littler, M. M., 1973. The population and community structure of Hawaiian fringing reef crustose Corallinaceae. J. Experiment. Mar. Biol. Ecol., 11:103-120.

Logan, B. W., 1961. *Cryptozoon* and associated stromatolites from the Recent, Shark Bay, Western Australia. J. Geol., 69:517-533.

Logan, B. W., Rezak, R. and Ginsburg, R. N., 1964. Classification and environmental significance of algal stromatolites. J. Geol., 72:68-83.

Logan, B. W., Hoffman, P. and Gebelein, C. D., 1974. Algal mats, cryptalgal fabrics, and structures, Hamelin Pool, Western Australia. In: Evolution and Diagenesis of Quaternary Carbonate Sequences, Shark Bay, Western Australia, Am. Assoc. Petroleum Geologists Mem. 22:140-194.

Lowenstam, H. A., 1955. Aragonite needles secreted by algae and some sedimentary implications. J. Sediment. Petrol., 25:270-272.

Lowenstam, H. A. and Epstein, S., 1957. On the origin of sedimentary aragonite needles of the Great Bahama Bank. J. Geol., 65:364-375.

Malan, S. P., 1964. Stromatolites and other algal structures at Mufulira, Northern Rhodesia. Econ. Geol., 59:397-415.

Mamet, B. and Roux, A., 1974. Sur quelques algues tubulaires scalariformes de la Téthys Paléozoique. Rev. Micropaléontol., 17(3):134-156.

Mamet, S. H., 1959. *Litostroma,* a new genus of problematical algae from the Pennsylvanian of Oklahoma. Am. J. Bot., 46(4):283-292.

Manten, A. A., 1971. Silurian reefs of Gotland. Elsevier, Amsterdam, 539 pp.

Marszalek, D. S., 1971. Skeletal ultrastructure of sediment producing green algae. In: Scanning Electron Microscopy/1971 (Pt. 1), Proc. 4th Ann. Scanning Electron Microscope Sympos., Chicago, 273-280.

Marszalek, D. S., 1975. Calcisphere ultrastructure and skeletal aragonite from the alga *Acetabularia antillana*. J. Sediment. Petrol., 45:266-271.

Maslov, V. P., 1956. Fossil calcareous algae of the USSR. Acad. Sci. USSR, Inst. Geol. Sci. Trudy, 160:301 pp.

Maslov, V. P., 1960. Stromatolites, their genesis, method of study, relationship with facies, and geological importance, based on the example of the Ordovician of the Siberian Plateau. Acad. Sci. USSR, Geol. Inst. Trudy, 41:188 pp.

Maslov, V. P., 1963. Introduction to the study of fossil Charophyta. Proc. G.I.N., Acad. Sci. USSR, 82:104 pp.

Maslov, V. P., et al., 1963. Algae. In: Y. A. Orlov (Editor), Basic Paleontology. Acad. Sci. USSR, Moscow, 14:19-312.

Maslov, V. P. and Kulik, Ye. L., 1956. New tribe of algae (Bereselleae) from the Carboniferous of the U.S.S.R. Acad. Sci. USSR, Doklady, 106:126-129.

Mason, L. R., 1953. The crustaceous coralline algae of the Pacific coast of the United States, Canada, and Alaska. Univ. Calif. Publ. Bot., 26(4): 313-390.

Massieux, M. and Denizot, M., 1964. Rapprochement due genre *Pseudolithothamnium* Pfender aves le genre actuel *Ethelia* Weber van Bosse (algues Florideae, Squamariaceae). Rev. Micropaléontol., 7(1):31-42.

Meijer, J. J. de, 1969. Fossil non-calcareous algae from insoluble residues of algal limestones. Leidse Geol. Med., 44:235-263.

Mendelsohn, F., 1973. Algae and ore deposits (abs.). Geol. Soc. Am. Abstr. Progr., 5(7):735.

Migula, W., 1897. Die Characeen Deutschlands, Oesterreichs und der Schweiz. In: L. Rabenhorst, Kryptogamen-Flora von Deutschland, Oesterreich und der Schweiz (2), Auflage 5. Leipzig, 765 pp.

Monty, C. L. V., 1967. Distribution and structure of Recent stromatolitic algal mats, eastern Andros Island, Bahamas. Ann. Soc. Géol. Belg., 90 (3):55-100.
Monty, C. L. V., 1971. An autoecological approach of intertidal and deep water stromatolites. Ann. Soc. Géol. Belg., 94:265-276.
Moore, C. H., Graham, E. A. and Land, L. S., 1976. Sediment transport and dispersal across the deep fore-reef and island slope (-55 m to -305 m), Discovery Bay, Jamaica. J. Sediment. Petrol., 46:174-187.
Munier-Chalmas, E., 1879. Observations sur les Algues calcaires confondues avec les Foraminifères et appartenant au groupe des Siphonées dichotomes. Soc. Géol. France Bull., ser. 3, 7:661-670.

Neumann, A. C. and Land, L. S., 1975. Lime mud deposition and calcareous algae in the Bight of Abaco, Bahamas. J. Sediment. Petrol., 45:763-786.
Nicholson, H. A. and Etheridge, R., 1878. A monograph of the Silurian fossils of the Girvan District in Ayrshire. Edinburgh and London, 1(1):135 pp.
Nitecki, M. H., 1970. North American cyclocrinitid algae. Fieldiana Geol., 21:1-182.
Nitecki, M. H., 1972. North American Silurian receptaculitid algae. Fieldiana Geol., 28:1-108.
Nitecki, M. H., 1972a. The paleogeographic significance of receptaculitids. Internat. Geol. Cong., 24th, Montreal, Proc. 7:303-309.

Ott, E., 1972. Mitteltriadische Riffe der Nörlichen Kalkalpen und altersgleiche Bildungen auf Karaburun und Chios (Agäis). Mitt. Ges. Geol. Bergbaustud., 21:251-276.
Ott, E., 1972a. Zur Kalkalgen-Stratigraphie der Alpinen Trias. Mitt. Ges. Geol. Bergbaustud., 21:455-464.
Owen, J. D., 1972. A log analysis method for Ekofisk Field, Norway. In: 13th Ann. Logging Symp., Soc. Prof. Well Log Anal. Trans., Paper X, 22 pp.

Park, R., 1976. A note on the significance of lamination in stromatolites. Sedimentology, 23:379-393.
Pautard, F. G. E., 1970. Calcification in unicellular organisms. In: H. Schraer, (Editor), Biological Calcification: Cellular and Molecular Aspects. Appleton-Century-Crofts, New York, 105-201.
Peck, R. E., 1957. North American Mesozoic Charophyta. U. S. Geol. Surv. Prof. Pap. 294-A, 44 pp.
Perkins, R. D., McKenzie, M. D. and Blackwelder, P. L., 1972. Aragonite crystals within codiacean algae: distinctive morphology and sedimentary implications. Science, 175:624-626.
Petryk, A. A. and Mamet, B. L., 1972. Lower Carboniferous algal microflora, southwestern Alberta. Can. J. Earth Sci., 9:767-802.
Philippi, R. A., 1837. Beweis das die Nulliporen Pflanzen sind. In: A. F. A. Wiegmann (Editor), Archiv. für Naturgeschichte. Berlin, 1(3): 387-393.
Pia, J., 1920. Die Siphoneae verticillatae vom Karbon bis zur Kreide. Abh. zool.-bot. Ges., Wien, 11(2):1-263.
Pia, J., 1926. Pflanzen als Gesteinbildner. Verlag von Gebrüder Borntraeger, Berlin, 355 pp.
Pia, J., 1927. Thallophyta. In: M. Hirmer (Editor), Handbuch der Paläobotanik, Oldenbourg, München, 1:31-136.
Pia, J., 1937. Die wichtigsten Kalkalgen des Jungpalaozoikums und ihre geologische Bedeutung. 2nd Cong. pour l'avancement des etudes de stratigraphie Carbonifere, (Heerlen, 1935), Compt. Rend., 2:765-856.
Pitcher, M., 1964. Evolution of Chazyan (Ordovician) reefs of eastern United States and Canada. Can. Petroleum Geol. Bull., 12:632-691.

Playford, P. E. and Cockbain, A. E., 1969. Algal stromatolites: deepwater forms in the Devonian of Western Australia. Science, 165:1008-1010.
Playford, P. E., Cockbain, A. E., Druce, E. C. and Wray, J. L., 1976. Devonian stromatolites from the Canning Basin, Western Australia. M. R. Walter (Editor), Stromatolites. Elsevier, Amsterdam, 543-564.
Pobeguin, T., 1954. Microstructure d'un algue calcaire: *Dactylopora* (Dasycladacée tertiaire): remarques sur les organismes aragonitiques et sur leur fossilization. Ann. Sci. Bot., Ser. 11, 15:325-336.
Pray, L. C. and Wray, J. L., 1963. Porous algal facies (Pennsylvanian), Honaker Trail, San Juan Canyon, Utah. In: R. O. Bass (Editor), Shelf Carbonates of the Paradox Basin. Four Corners Geol. Soc. Sympos., 4th Field Conf., 204-234.

Rezak, R., 1959. New Silurian Dasycladaceae from the southwestern United States. Colorado School Mines Quart., 54(1):115-129.
Rezak, R., 1971. Reproduction and growth rates. In: R. Ginsburg, R. Rezak and J. L. Wray, Geology of Calcareous Algae (Notes for a Short Course). Comparative Sediment. Lab., Univ. Miami, 3.1-3.8.
Rich, M., 1967. *Donezella* and *Dvinella*, widespread algae in Lower and Middle Pennsylvanian rocks in east-central Nevada and west-central Utah. J. Paleontol., 41:973-980.
Riding, R., 1974. The Devonian genus *Keega* (Algae) reinterpreted as stromatoporoid basal layer. Palaeontology, 17:565-577.
Riding, R., 1975. *Girvanella* and other algae as depth indicators. Lethaia, 8:173-179.
Riding, R. and Brasier, M., 1975. Earliest calcareous foraminfera. Nature, 257(5523):208-210.
Riding, R. and Jansa, L. F., 1974. *Uraloporella* Korde in the Devonian of Alberta. Can. J. Earth Sci., 11:1414-1426.
Riding, R. and Wray, J. L., 1972. Note on the ?algal genera *Epiphyton*, *Paraepiphyton*, *Tharama* and *Chabakovia*. J. Paleontol., 46:918-919.
Rietschel, S., 1969. Die Receptaculiten. Eine Studie zur Morphologie, Organisation, Ökologie und Überlieferung einer problematischen Fossil-Gruppe und die Deutung ihrer Stellung im System. Senck. Lethaea, 50: 465-517.
Rothpletz, A., 1891. Fossile Kalkalgen aus den Familien der Codiaceen und der Corallineen. Zeitschr. Deutsch. Geol. Ges., 43:295-322.
Rupp, A. W., 1967. Origin, structure, and environmental significance of Recent and fossil calcispheres (abs.). Geol. Soc. Am. Spec. Pap., 101:186.

Schopf, J. W., 1968. Microflora of the Bitter Springs Formation, late Precambrian, cental Australia. J. Paleontol., 42:651-688.
Schopf, J. W., Oehler, D. Z., Horodyski, R. J. and Kvenvolden, K. A., 1971. Biogenicity and significance of the oldest known stromatolites. J. Paleontol., 45:477-485.
Schroeder, J. H., 1972. Calcified filaments of an endolithic alga in Recent Bermuda reefs. N. Jb. Geol. Paläont. Mh., 1972(1):16-33.
Setchell, W. A., 1905. Post-embyonal stages of the Laminariaceae. Univ. Calif. Publ. Bot., 2:115-138.
Seward, A. C., 1898. Fossil Plants. University Press, Cambridge, 1:452 pp.
Sloane, H., 1707. A Voyage to the Islands Madera, Barbados, Nieves, S. Cristophers and Jamaica. London, 1:264 pp.
Soegiarto, A., 1973. Benthic algae of the bay. In: S. V. Smith, K. E. Chave and D. T. O. Kam (Editors), Atlas of Kaneohe Bay - A Reef Ecosystem Under Stress. University of Hawaii, UNIHI-SEAGRANT-TR-72-01, 67-90.
Stanton, R. J., 1963. Upper Devonian calcispheres from Redwater and South Sturgeon Lake reefs, Alberta, Canada. Can. Petroleum Geol. Bull., 11(4): 410-418.

Stieglitz, R. D., 1973. Carbonate needles: additional organic sources. Geol. Soc. Am. Bull., 84:927-930.
Stockman, K. W., Ginsburg, R. N. and Shinn, E. A., 1967. The production of lime mud by algae in south Florida. J. Sediment. Petrol., 37:633-648.
Stolley, E., 1893. Ueber silurische Siphoneen. N. Jb. Min., 2:135-146.
Surdam, R. C. and Wray, J. L., 1976. Lacustrine stromatolites, Eocene Green River Formation, Wyoming. In: M. R. Walter (Editor), Stromatolites. Elsevier, Amsterdam, 535-542.
Szulczewski, M., 1963. Stromatolites from the high-tatric Bathonian of the Tatra Mountains. Acta Geol. Polon., 13:142-145.

Termier, H., Termier, G. and Vachard, D., 1975. Sur la systematique et la phylogenie des Moravamminida et des Aoujgaliia (abs.): Internat. Sympos. Fossil Algae, Abstr. Progr., Erlangen, October, 1975:40.
Terry, C. E. and Williams, J. J., 1969. The Idris "A" bioherm and oilfield, Sirte Basin, Libya -- its commercial development, regional Paleocene geologic setting and stratigraphy. In: P. Hepple (Editor), The Exploration for Petroleum in Europe and North Africa. Inst. Petroleum, London, 31-48.
Toomey, D. F. and Klement, K. W., 1966. A problematical micro-organism from the El Paso Group (Lower Ordovician) of west Texas. J. Paleontol., 40:1304-1311.

Unger, F., 1858. Beiträge zue näheren Kenntniss des Leithakalkes. Denkschr. K. Akad. Wiss. München, 14:13-38.

Valet, G., 1968. Contribution à l'étude des Dasycladales - Morphogenèse. Nova Hedwigia, 16:21-82.
Valet, G., 1969. Contribution à l'étude des Dasycladales - Cytologie et reproduction, révision systématique. Nova Hedwigia, 17:551-644.

Walcott, C. D., 1914. Pre-Cambrian Algonkian algal flora. Smithsonian Inst. Misc. Coll., 64(2):77-156.
Walter, M. R., 1972. Stromatolites and the biostratigraphy of the Australian Precambrian. Palaeontol. Assoc. Spec. Publ. 11:191 pp.
Walter, M. R., (Editor), 1976. Stromatolites: Elsevier, Amsterdam, 772+ pp.
Wilbur, K. M., Colinvaux, L. H. and Watabe, N., 1969. Electron microscope study of calcification in the alga *Halimeda* (order Siphonales). Phycologia, 8:27-35.
Wilbur, K. M. and Watabe, N., 1963. Experimental studies on calcification in molluscs and the alga *Coccolithus huxleyi*. Ann. N. Y. Acad. Sci., 109:82-112.
Wilson, E. C., 1969. No new *Ungdarella* (Rhodophycophyta) in New Mexico. J. Paleontol., 43:1245-1247.
Wilson, E. C., Waines, R. H. and Coogan, A. H., 1963. A new species of *Komia* Korde and the systematic position of the genus. Palaeontology, 6:246-253.
Wiman, S. K. and McKendree, W. G., 1975. Distribution of *Halimeda* plants and sediments on and around a patch reef near Old Rhodes Key, Florida. J. Sediment. Petrol., 45:415-421.
Wincander, E. R. and Schopf, J. W., 1974. Microorganisms from the Kalkberg Limestone (Lower Devonian) of New York state. J. Paleontol., 48:74-77.
Wood, A., 1941. "Algal dust" and the finer-grained varieties of Carboniferous limestones. Geol. Mag., 78(3):192-200.
Wood, R. D. and Imahori, K., 1964. Iconograph of the Characeae. In: R. D. Wood and K. Imahori (Editors), A Revision of the Characeae, 2. Verlag Von J. Cramer, Weinheim, pls. 1-395.

Wood, R. D. and Imahori, K., 1965. Monograph of the Characeae. In: R. D. Wood and K. Imahori (Editors), A Revision of the Characeae, 1. Verlag Von J. Cramer, Weinheim, pp. 1-904.

Wray, J. L., 1964. *Archaeolithophyllum*, an abundant calcareous alga in limestones of the Lansing Group (Pennsylvanian), southeastern Kansas. Kansas Geol. Surv. Bull., 170(1):1-13.

Wray, J. L., 1967. Upper Devonian calcareous algae from the Canning Basin, Western Australia. Colorado School Mines Prof. Contrib., 3:76 pp.

Wray, J. L., 1968. Late Paleozoic phylloid algal limestones in the United States. Internat. Geol. Congr., 23rd, Prague, Proc. 8:113-119.

Wray, J. L., 1969. Paleocene calcareous algae from Libya (abs.). Symposium on Geology of Libya, University of Libya, Prog. Abstr., Tripoli, April, 1969:21-22.

Wray, J. L., 1971. Ecology and geologic distribution. In: R. Ginsburg, R. Rezak and J. L. Wray, Geology of Calcareous Algae (Notes for a Short Course). Comparative Sediment. Lab., Univ. Miami, 5.1-5.6.

Wray, J. L., 1972. Environmental distribution of calcareous algae in Upper Devonian reef complexes. Geol. Rundsch., 61:578-584.

Wray, J. L., 1972a. Submarine reef exploration in British Honduras. Colorado Mines Mag., 62(4):19-21.

Wray, J. L., James, N. P. and Ginsburg, R. N., 1975. The puzzling Paleozoic phylloid algae -- Holocene answer in squamariacean calcareous red algae (abs.). Am. Assoc. Petroleum Geologists, Ann. Mtgs. Abstr., 2:82-83.

Wray, J. L. and Playford, P. E., 1970. Some occurrences of Devonian reef-building algae in Alberta. Can. Petroleum Geol. Bull., 18:544-555.

Young, G. M. and Long, D. G. F., 1976. Stromatolites and basin analysis: an example from the upper Proterozoic of northwestern Canada. Palaeogeogr. Palaeoclimatol. Palaeoecol., 19:303-318.

Zankl, H., 1971. Upper Triassic carbonate facies in the northern Limestone Alps. In: G. Müller (Editor), Sedimentology of Parts of Central Europe. Guidebook, VIII Internat. Sedimentol. Congr., Heidelberg, 147-185.

INDEX